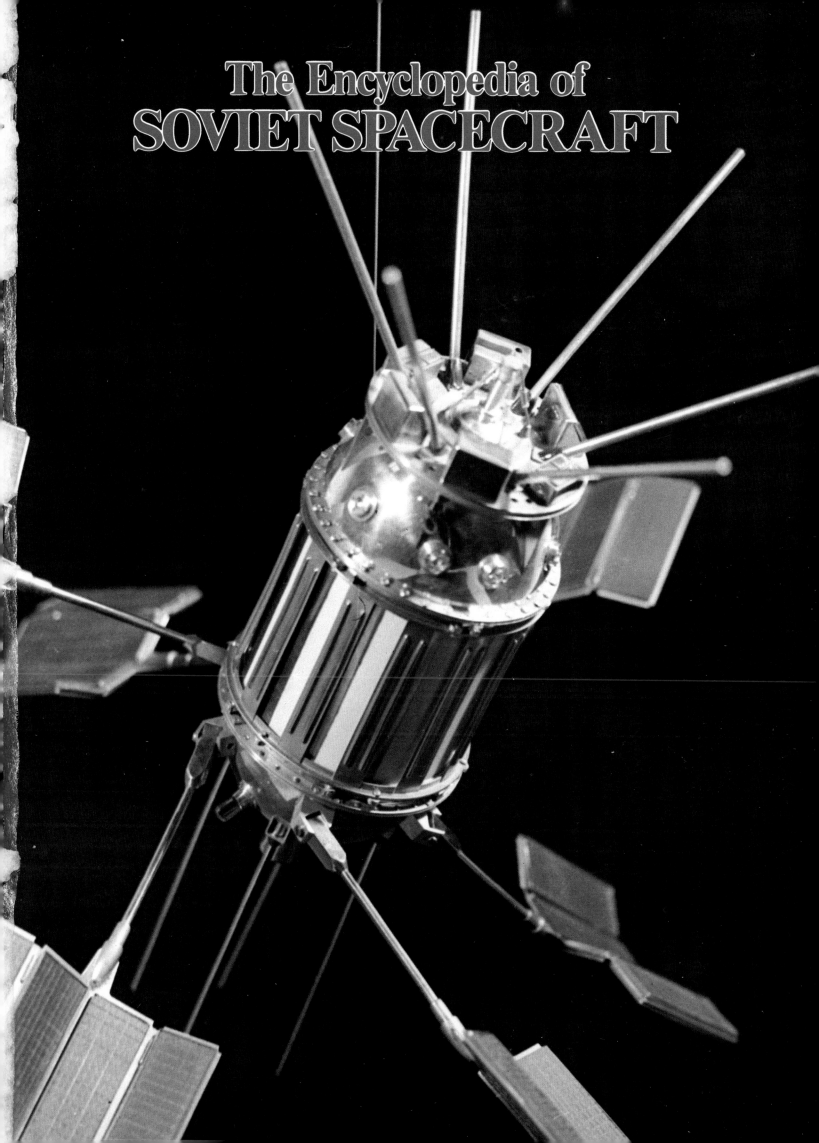

The Encyclopedia of
SOVIET SPACECRAFT

The Encyclopedia of
SOVIET SPACECRAFT

Douglas Hart

Exeter Books

NEW YORK

A Bison Book

ENCY OF SOVIET SPACECRAFT

First published in USA 1987
by Exeter Books
Distributed by Bookthrift
Exeter is a trademark of Bookthrift.
Marketing, Inc.
Bookthrift is a registered trademark of
Bookthrift Marketing
New York, New York

ISBN 0-671-08932-3

Printed in Hong Kong

Picture Credits

Jan Cazden: 7, 16, 18 (left)
Department of Defense: 180 (left), 184 (bottom left)
R F Gibbons: 17 (bottom), 36,39, 41, 52-53, 82, 87 (left), 88, 89 (bottom), 90, 93, 96 (bottom), 97, 98-99, 100 (top), 101 (top), 102-103 (bottom), 104, 106-107, 108-109, 148, 151 (top right), 156-157 (both), 158, 164, 169, 175
NASA: 86, 94-95, 98 (top), 102 (top)

Novosti: 1, 2-3, 8, 9 (both), 11, 14-15 (both), 20, 22-23, 24, 25, 26 28, 29, 33, 45, 52 (top left), 56-57 (all), 58-59, 60, 61, 62, 65 (right), 67 (top left), 68, 72-73, 76, 77 (top right and bottom), 79, 83, 92 (bottom), 96, 102-103 (top), 109 (top and bottom right), 110 (left), 112, 113, 116, 117, 124 (top), 127 (top left), 130, 131, 134 (top), 135, 138-139, 142-143, 144, 145, 146-147, 149, 150-151, 152, 153, 154-155, 155 (top and bottom right), 159, 161, 162 (bottom), 165, 166-167, 168, 170-171 (both), 172-173, 173 (right), 177, 178, 180-181, 182, 183 (right), 184 (top left), 184-185, 191 (right)
Smithsonian Institution: 47, 52 (bottom), 60, 63, 64, 67 (bottom left), 69 (bottom), 70, 71, 77 (left), 78, 122 (bottom left), 124 (bottom), 127 (top right and bottom), 132, 133, 134 (bottom), 136 (top right)
Society for Cultural Relations with the USSR: 65 (left), 89 (top), 120, 121, 125, 128-129, 136 (top left and bottom), 183 (top left), 191 (left)
Soviet ASTP Press Kit: 95 (right), 100 (bottom), 101 (bottom)
TASS from Sovfoto: 4-5, 12-13, 18-19, 27, 30-31, 36-37, 69 (top), 81, 105, 110-111, 114-115 (both), 118-119, 122-123 (both), 126, 140-141, 162-163, 183 (bottom left), 186-187, 189, 192
G H Thomson: 151 (bottom right)
US Department of Defense: 17 (top), 40, 55, 84, 174, 180 (left), 184 (bottom left)
USSR Academy of Sciences (via NASA): 130 (left)
D R Woods: 66, 74, 137, 176
© **Bill Yenne:** 87 (right), 107, 160, 187 (right)

Acknowledgements
We wish to thank the following people for helping to make this book possible: Ralph F Gibbons of the Kettering Group; Barbara Lewis, our translator; Paul Stares of the Brookings Institution; with a special thank-you to Wendy Sacks of Bison Books Ltd in London.

Edited by John Kirk; designed by Ruth DeJauregui; picture research by Bill Yenne and Ruth De-Jauregui; and captioning by Timothy Jacobs.

Page 1: The Elektron 1 satellite was, with its companion Elektron 2, the Soviets' first dual vehicle launch. Still early in the Soviet space program, such a multiple 'package' may have presaged the complex Soviet interplanetary probes of later years.

Page 2–3: At the eastern end of the Baikonur Cosmodrome, a large D-1e launch vehicle is being transported to its gantry. The D-1e is used for launching Luna, Zond, Mars and Venera spacecraft. Its first stage develops 1,496,880 kg of thrust.

Below: Near Dzhezkazgan, recovery crewmembers gather around the Soyuz T-12 descent module after the craft's historic flight—Svetlana Savitskaya's three-hour and 35 minute EVA outside of the Salyut 7 space station. *Note* the men kneeling are about to open the capsule's hatch, thus welcoming Svetlana and her crewmates Vladimir Dzhanibekov and Igor Volk to the familiar air of the USSR.

Contents

INTRODUCTION

Russian interest in space exploration predates the Bolshevik Revolution. At the turn of the century Konstantin Tsiolkovsky, the intellectual patriarch of the USSR's space program, wrote about establishing human colonies in near-Earth orbit as a prelude to manned investigation of the planets, and eventually of the stars. According to Tsiolkovsky, mankind would never be free to evolve toward perfection until it had conquered the artificial limits imposed upon it by gravity. This visionary attitude continues to influence the Soviet space effort. In the Soviet Union it has long been assumed as a matter of course that mankind will eventually spread beyond the planet of its origin and that Soviet citizens will be in the vanguard of this effort.

The triumph of Marxist-Leninism and the establishment of what space historian Walter McDougall terms the first technocratic state transformed Russian from a feudal autocracy, where men like Tsiolkovsky were relegated to dreams and the written word, to a modern autarchy capable of harnessing the resources of the world's largest country for the purposes of technical advancement. In the end, however, it remained men of extraordinary talent who launched the space age in the USSR and the world at large. Men like Sergei Korolev and Vladimir Chelomei overcame the obstacles of state terrorism, the Nazi invasion and the entrenched official paranoia of the initial stages of the cold war to develop the technology that made the USSR the first nation into space.

The Soviet leadership has sought to exploit the political and propaganda impact of its space program from its very inception. This has proven a mixed blessing for the USSR, since it led to an early requirement for increasingly spectacular 'space firsts' which in the long run damaged the Soviets vis-a-vis their great rival in space, the United States. The early Sputniks, the first manned space launches and the perception of a race to the moon (fueled as much by the East as by the West, as long as it was to Moscow's advantage) tended to obscure the more basic difference between the Russian and the American approaches to space exploration and exploitation. While the US tends to de-

Left: Important communications—this montage conveys the complexity of operations at the Soviet flight control center during the Soyuz-Apollo joint mission in 1975. *Above:* A Soviet launch vehicle for a Vostok mission on display.

sign programs and technology to produce quantum leaps in capability between generations, the USSR tends to rely on a more incremental approach, in which proven technology is slightly modified over a series of generations. The American approach has produced periodic 'recessions' in the US space program while new goals and technology were being developed. This was the case with American manned space efforts between the Apollo and the space shuttle, and is currently the case with respect to the space station successor to Sky Lab, which is not expected until the 1990s. Throughout this period the Soviets have continued to press on, and though their technology has been clearly inferior, their man-hours in orbit, experience with materials processing in space and expertise in space medicine now far exceed that of the US.

The early years of the Soviet space program were spent in direct competition with the US. As long as the USSR could appear technically superior to the Americans, the leadership pushed for comparison. But the Soviets were forced to abandon this approach in the 1960s as the US marshalled its technical resources. First, Nikita Khrushchev's minatory strategic rhetoric, based almost entirely on the early 'space spectaculars,' was revealed to be groundless during the Cuban missile crisis. Second, it was clear by the latter part of the decade that the US was going to be the first to place a man on the moon. The Soviets characteristically opted out of a race they could not win. Instead, during the 1970s, they maximized the benefits to be gained from their incremental approach while political and budgetary malaise afflicted the United Sates in the aftermath of the Vietnam War.

While civilian space programs were receiving the most attention during the 1960s and 1970s, the US and the USSR steadily increased the resources they applied to the military exploitation of space. Competition has never slackened in this arena. From the end of the Second World War the US has been openly interested in satellite recon-

On Alpha Ralpha Boulevard: Cordwainer Smith's fictional skyway intersects with Bulgarian artist O Yankov's 'The Path to the Stars.' The far-off nebula looks a bit like a winged man —Daedalus, not Icarus, one assumes.

naissance as a means of piercing the veil of secrecy surrounding the USSR. Although the Soviet Union publicly denounced spy satellites, it secretly developed them, and, since the debut of Kosmos 4 in 1962, photo-reconnaissance missions have constituted the vast majority of Soviet space launches. Inexorably, both superpowers moved beyond reconnaissance to communications, early warning and weapons applications in space. Although weapons of mass destruction have been banned from orbit by treaty, both sides have developed anti-satellite weapons and appear to be poised on the verge of expanding the competition to orbital anti-missile systems.

The 1980s have brought major advances and serious setbacks to both the US and Soviet space programs. The American return to manned space flight, with a reuseable space transportation system, has been marred by the Challenger disaster and several major difficulties with civilian and military unmanned launch vehicles. The Soviets have neither chalked up such impressive gains nor suffered such devastating catastrophes. The placement of the new third generation Mir space station into low Earth orbit illustrates current Soviets strengths and weaknesses in space. Mir probably represents the penultimate step in the long-standing Soviet goal of large, manned, permanent space stations. Yet the vehicle itself was placed in orbit by a booster that is 15 years old. Soviet failure to develop a heavy Saturn V-type launch vehicle or a reusable spacecraft has certainly inhibited the USSR's space program, and the balance of the decade should prove interesting, as the US struggles to overcome its current setbacks, and the USSR attempts to deploy technology the lack of which has constrained its space program since the mid-1960s.

Soviet Spacecraft

Although the Soviet Union has of late become more open concerning its space program (*eg,* allowing the live broadcast of a space launch in early 1986), little can be found in the open Western literature concerning the civilian/military division of effort or the facilities and organizations that manufacture and assemble Soviet spacecraft. Unlike the US space program, there does not appear to be any clear demarcation between civilian efforts and those controlled by the military. While it is clear from mission type that well over 50 percent of Soviet space launches are devoted to purely military purposes, it is less clear what additional military applications may be entailed in ostensibly civilian missions. All launches are conducted by the Soviet

Strategic Rocket Forces, but about 10 minutes into flight they are handed over to various control centers depending upon mission type. The location and some of the functions of these control centers are known and discussed below, but their affiliation and subordination are less clear. In 1984 the US Defense Intelligence Agency published an organizational diagram describing the relationship between various Soviet government and party agencies involved in the space program. This diagram is reproduced

Below: Astronauts Popov and Ryumin in the Gagarin training center planetarium. *Bottom:* This Arkangelsk station monitors Northern USSR and Indian Ocean area test balloons.

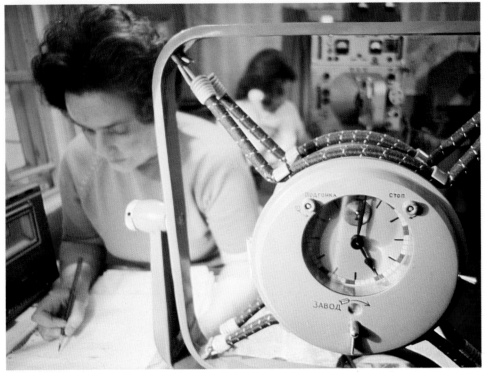

in Figure 1. What the diagram does not include, however, is Glavkosmos SSR, or the Main Administration for the Development and Use of Space Technology for the National Economy and Scientific Research. Founded in 1985, Glavkosmos is charged with coordinating Soviet civilian space activities, launches and cooperative international missions. Glavkosmos was evidently created to fill an organizational gap in the direction of the Soviet space program. While most Western space powers long ago formed central organizations, such as NASA in the United States, to coordinate their civilian space programs, the Soviets evidently never established an entity that would play a coordinating role between laboratories and aerospace design bureaus, as well as managing cooperative international programs. Perhaps more to the point, however, the relationship of Glavkosmos to military organizations within the Soviet space program, and the percentage of the estimated $20–25 billion annual space budget actually controlled by Glavkosmos, are unknown.

Even less information is available on manufacturing and assembly facilities for Soviet launch vehicles and spacecraft. Locations have never been mentioned in the open literature, and pictures of Soviet assembly lines for space craft and rockets are extremely rare. It is thought that many Soviet satellites are configured from a series of common external shells, allowing a modular assembly process to occur at sites at or near the three major cosmodromes. Alternatively, partially or fully assembled spacecraft may be shipped to their launch sites on the Soviet Union's extensive rail network.

Soviet Launch Sites and Ground Support Facilities

The Soviet Union has three principal cosmodromes: Tyuratam, Plesetsk and Kapustin Yar. Tyuratam is located in Kazakhstan around 45.6N/63.4E. The Soviets still refer to this sprawling launch complex as the Baykonur Cosmodrome, an old deception designed to mislead the West by referring to a town some 370 kilometers distant from the city of Leninsk, which has grown up around the railhead near the launch site. The first Soviet ICBM test launches were from Tyuratam, followed shortly by the Sputnik and manned Vostok missions that inaugurated the space age. Tyuratam continues to launch both civilian and military missions and remains the exclusive launch site for manned and interplanetary flights. In addition, most experimental and developmental space efforts are directed from Tyuratam. In the past such programs have included the fractional orbital bombardment system, anti-satellite interceptors and all new manned spacecraft and space stations. Systems currently under development at Tyuratam include the Soviet version of the space shuttle and new family of medium and heavy lift boosters.

In 1966 the USSR began launches from its northernmost complex near Plesetsk (62.8N/40.1E). This cosmodrome is the primary spaceport for both the Soviet military and all navigation and meteorological satellite launches. The USSR has never officially acknowledged the presence of a launch complex at Plesetsk, despite the fact that it has boasted the highest annual launch rate of any launch facility in the world for almost two decades.

Since 1962, small and intermediate-size Soviet payloads have been orbited from Kapustin Yar, located near Volgograd at 48.4N/45.8E. Although the range is apparently used primarily for testing medium-range military missiles, the Soviets have been the most open with access to this launch complex. Engineers from the Warsaw Pact nations, France, India and Sweden have all been allowed to view launches at Kapus-

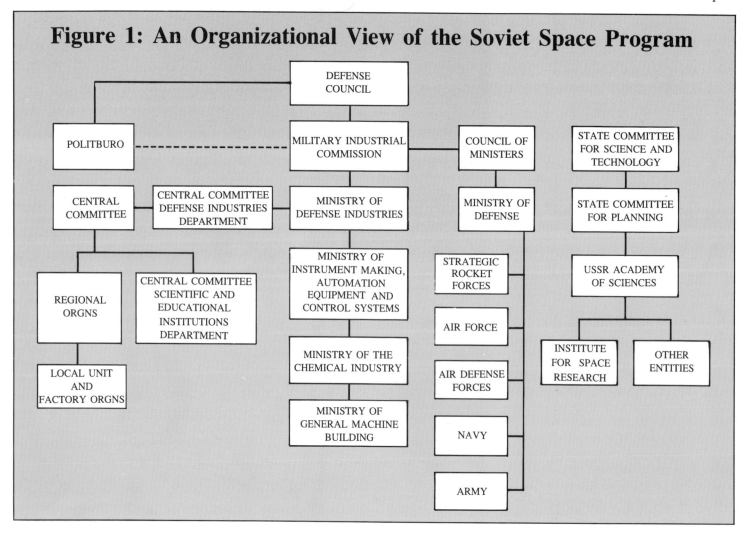

Figure 1: An Organizational View of the Soviet Space Program

tin Yar. The most important missions launched recently from this site have been sub-scale models of a Soviet spaceplane.

Together, the three Soviet cosmodromes support an impressive launch schedule. The Soviets conduct a space launch once every four days. Two launches are often conducted within 24 hours from one or two cosmodromes. On occasion the USSR has launched three spacecraft within an eight-hour period. The Soviets are also capable of sustaining high launch rates over a week's time.

The reader should be cautioned that the large number of annual Soviet launches (98 in 1985, as compared to 17 by the US), is partially a product of inferior technology and higher failure rates than equivalent Western systems. These massive launch rates do, however, serve as a valuable experiential base for a more mature space program in the next century, and they imply an impressive capability for replacing lost or damaged military space assets during a conflict.

Soviet spacecraft are monitored by a series of ground- and ocean-based stations using radar, laser and optical technology. There are seven major tracking stations located in the USSR: Yevpatoria, Tbilisi, Dzhusaly, Kolpashevo, Ulan Ude, Ussuriysk and Petropavlovsk. Additional tracking stations are located in Eastern Europe, Cuba, Africa, the southern Indian Ocean and Antarctica. The land-based facilities are supported by a fleet of ships equipped with tracking equipment. These ships belong to three basic classes: the *Kosmonaut Yuri Gagarin,* the *Kosmosnaut Vladimir Komarov* and the *Akademic Sergei Korolov.* In addition to civilian facilities, Soviet ballistic missile warning radars support space tracking/surveillance efforts. The primary components of this network are the 11 HEN HOUSE radars deployed around the Soviet periphery, and the six large phased array radar complexes being built to replace the HEN HOUSE systems.

Control facilities for Soviet space missions have been located at Kaliningrad, Yevpatoria and Simeiz. Kaliningrad is usually referred to as controlling manned missions, while interplanetary flights are handled from the latter two stations. Military control centers doubtless exist, but their locations have yet to come to light in the available open literature.

The Soviets maintain a large and modern research, test and training infrastructure for their space program, but precise details concerning locations, staffs and resource alloca-

Below: **This Soviet 8-dish antenna, probably part of the Tbilisi facility, scans Crimean skies for civilian and military spacecraft transmissions.**

tions are scarce. Early rocket engines were developed at the Leningrad Gas Dynamics Laboratory, but the status of current R&D facilities remains sketchy. The only specific organizations known to be dedicated to space science are the Moscow Space Research Institute of the Soviet Academy of Sciences and the Vernadsky Institute of Geochemistry and Analytical Chemistry. The Space Research Institute maintains a major complex in the Moscow area and has branches in other parts of the Soviet Union. The Vernadsky Institute has a staff of around 1200 persons distributed throughout 30 laboratories. The major test and training center for Soviet space projects is at Star City (Zvezdnyy Gorodok) near Moscow. At this facility exact replicas of spacecraft are maintained in conditions as similar as possible to those experienced by operational vehicles, so that solutions to problems can be attempted on the ground before a possibly fateful decision must be taken regarding the operational systems. Star City is also home to the Yuri Gagarin Cosmonaut Training Center. Additional cosmonaut training facilities are also located at Leninsk. These installations are similar to those run by NASA at the Johnson Space Center in Houston, Texas, and the Kennedy Space Center at Cape Canaveral, Florida.

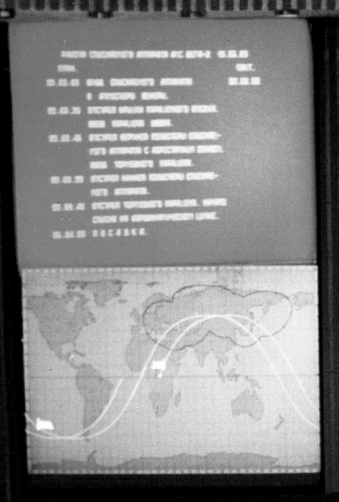

Earth to Venus: Seen *here* is a portion of the flight control center for the Soviet Venus encounter of 15 June 1985. *To the left* is a chart of the vehicle's ground traces while in Earth orbit and *at center* is a sequential chart featuring the baloon-borne probe and a lander unit.

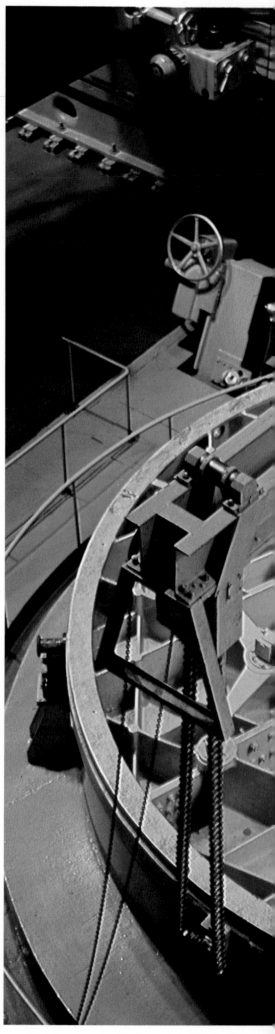

This work provides general reference material concerning the civilian and military spacecraft of the Union of Soviet Socialist Republics. The ambitious Soviet launch schedule (currently around 100 launchings a year) and pervasive secrecy concerning all aspects of the USSR's space program make it impossible to supply physical and mission data on every spacecraft that the Soviets have launched, or attempted to launch, into orbit. It is possible, however, to describe the salient aspects of some 37 Soviet spacecraft programs by drawing on certain authoritative Western sources, and within each program category individual vehicles and their systems can be described.

It has been Soviet practice from the inception of the USSR's space program both to capitalize on the propaganda value of the effort's successes and to minimize any information concerning program failures. Thus the availability of data concerning Soviet spacecraft can be organized hierarchically in the following manner: civilian/scientific missions that the Soviets consider successful; civilian/scientific missions that are partially successful; civilian/scientific missions that are complete failures; and all military missions. Indeed, although the great majority of Soviet space launches are partially or completely devoted to military missions, the USSR has never acknowledged any military role for its space program. While Soviet civilian/scientific spacecraft are of necessity treated in greater detail in this encyclopedia, what data that is available concerning Soviet military spacecraft has been compiled from open-source United States government publications, trade journals and other specialist publications and the Western press.

Above: **Soyuz spacecraft in the assembly shop. Though control failure caused Soyuz 1 to crash, various versions of this craft have been the cosmonauts' 'Old Reliable'.**

It would not be possible to discuss any spacecraft independently from either the vehicles used to place them in orbit or the ground-based infrastructure designed to support space missions. This is especially the case with the Soviet space program, given its emphasis to date upon large numbers of space launches utilizing expendable boosters, and the virtually complete control of all missions that has been exercised by ground-based controllers (as opposed to the crews piloting the craft) since the very first Sputnik was launched in 1957. Soviet launch sites and ground support facilities are discussed in the introduction to this encyclopedia. Soviet space boosters are examined in detail in a separate appendix.

The Future of Soviet Spacecraft

On 19 February 1986 the Soviet Union successfully launched the initial component of what Moscow hopes will become the first permanently manned earth-orbiting space station. Such an accomplishment has been a long-standing goal of the Soviet space program. Christened Mir, or Peace, this follow-on to the Salyut series of manned space stations clearly demonstrated the major Soviet commitment to manned space missions. Although the launch of the Mir space station was obviously planned long in advance, the fact of its coming so soon after the catastrophic launch failure of the US space shuttle Challenger

Seen in the round: the Zelenchuk spherical astrophysical observatory at the Soviet Academy of Science, on Profcoyuznaya Vlitza in Moscow.

In keeping with the Soviet veneration of industry and technology, this monument to the space program is done in the Futurist style with which Malevich, Tatlin and others celebrated the Revolution in the teens and twenties of this century. Far from the lights of ground, swept upward on its lines of thrust, the spacecraft at the top of this monument is evidenced by its barely-visible tailfin, *at center of photo*. The monument is in Moscow, near the Economic Achievements Exhibition.

had the effect of demonstrating the propaganda value of the Soviet space program, which every Soviet leader since Nikita Khruschev has sought to exploit to the USSR's advantage. (A separate appendix to this volume details the Soviet space stations that have been launched to date.)

The Soviet space program has been characterized by steady incremental growth in resource allocation and program objectives since the advent of Leonid Brezhnev's regime in the early 1960s. Much like the Soviet military build-up, which has received so much attention in the West, the Soviet space program has emphasized sustained growth through a succession of modest objectives, with a view toward the long-term aspects of competition with the capitalist democracies.

There are some reasons to believe, however, that Soviet resources devoted to space could increase dramatically under the tenure of the new Soviet leadership headed by Mikhail Gorbachev. The new Party Secretary finds himself in a position not unlike that of John Kennedy, replacing an aging, cautious leadership and with a mandate to 'get the country moving again.' Like Kennedy, Gorbachev could well come to see an accelerated space program as a means of differentiating his administration from those of the past, firing the popular imagination and competing successfully with adversaries

abroad. Chairman Gorbachev can look forward to perhaps two decades of uninterrupted rule if he remains physically and politically healthy—a luxury possessed by no democratic regime, and one which lends itself to the long-range planning that is essential to a successful space program.

In any event, it is clear that the Soviets will continue to support a continuing effort with respect to space applications and exploration, and the general trends of this program can be projected into the early twenty-first century. A permanent manned space station will allow the Soviets to continue to expand their experiments with manufacturing in space, with an eventual goal of establishing entire industries in orbit. Near-term efforts will concentrate upon manufacturing processes that are difficult or impossible outside a zero-gravity environment, as well as experiments in collecting solar power for dissemination back to earth or to other platforms in orbit.

Examples of such planned Soviet space applications efforts include 'Star Electricity' and a project of the Moscow Institute of Avionics concerning the nighttime illumination of large terrestrial areas. Announced in early 1985, 'Star Electricity' is an attempt to construct a power plant in near-Earth orbit that would beam collected solar power back to Earth using super-high frequencies. The

Upper right: A Soviet dream is portrayed in this artist's vision of a Soviet space shuttle, approaching a multi-docking port Soviet Mir space station.

nighttime illumination effort should be initiated in the near future by orbiting an experimental 200-kilogram test reflector. Soviet specialists have projected an operational network of these umbrella-shaped reflectors by the 1990s, designed to provide inexpensive nighttime lighting for cities, road and rail networks and large construction projects in remote northern locations of the USSR where daylight is scarce during the winter months.

The Soviets should also be able to deploy a reusable space transportation system before the turn of the century. Two types of orbiters have been associated with this effort: a reusable spaceplane capable of carrying a multi-member crew and a small payload, and a vehicle closely resembling the US space shuttle and capable of orbiting significant payloads.

Planetary exploration will also be another priority in the civilian part of the Soviet space program. Probes will continue to be sent to Venus and Mars, and a manned mission to one of the planets (probably Mars) is a distinct possibility in the early twenty-first century—perhaps in conjunction with the centennial celebration in 2017. Sometime in

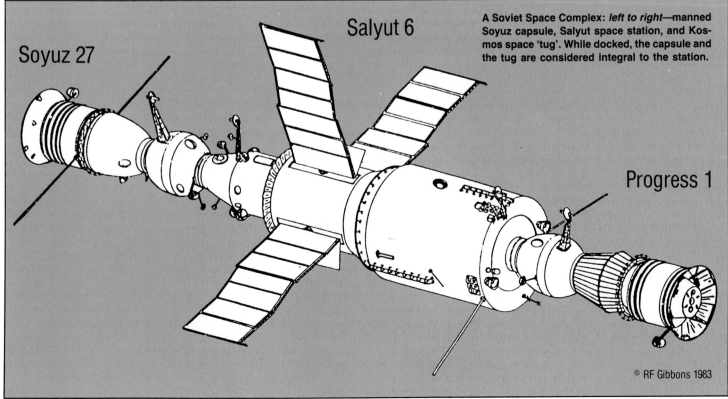

Soyuz 27

Salyut 6

Progress 1

A Soviet Space Complex: *left to right*—manned Soyuz capsule, Salyut space station, and Kosmos space 'tug'. While docked, the capsule and the tug are considered integral to the station.

© RF Gibbons 1983

that century mankind will launch its first probes to another star system. Whether the USSR will seek the propaganda advantage of being the first to attempt such an undertaking or will join in an international effort remains to be seen.

Military applications will continue to receive the bulk of Soviet resources dedicated to the space program. Advances will be sought in the areas of intelligence collection, weapons targeting, command and control, communications, weather forecasting and geodesy. Research and development concerning space weaponry for anti-satellite, anti-ballistic missile and space-to-ground delivery applications could continue at current or accelerated levels, depending upon the

Above: Yuri Gagarin smiles from a photo at the Economic Achievements Exhibition, in Moscow. *Right:* The K Tsiolkovsky State Museum.

outcome of arms control negotiations with the United States and upon the results of the American Strategic Defense Initiative.

In speculating about the future scope and direction of the Soviet space effort it is important to keep the past firmly in mind. American and German contributions have advanced the field, but pride of place in both rocketry and spaceflight clearly belong to Russian pioneers. The country that developed the first intercontinental ballistic missile, launched the first artificial satellite, and put the first man into space could still have a few surprises in store for the West.

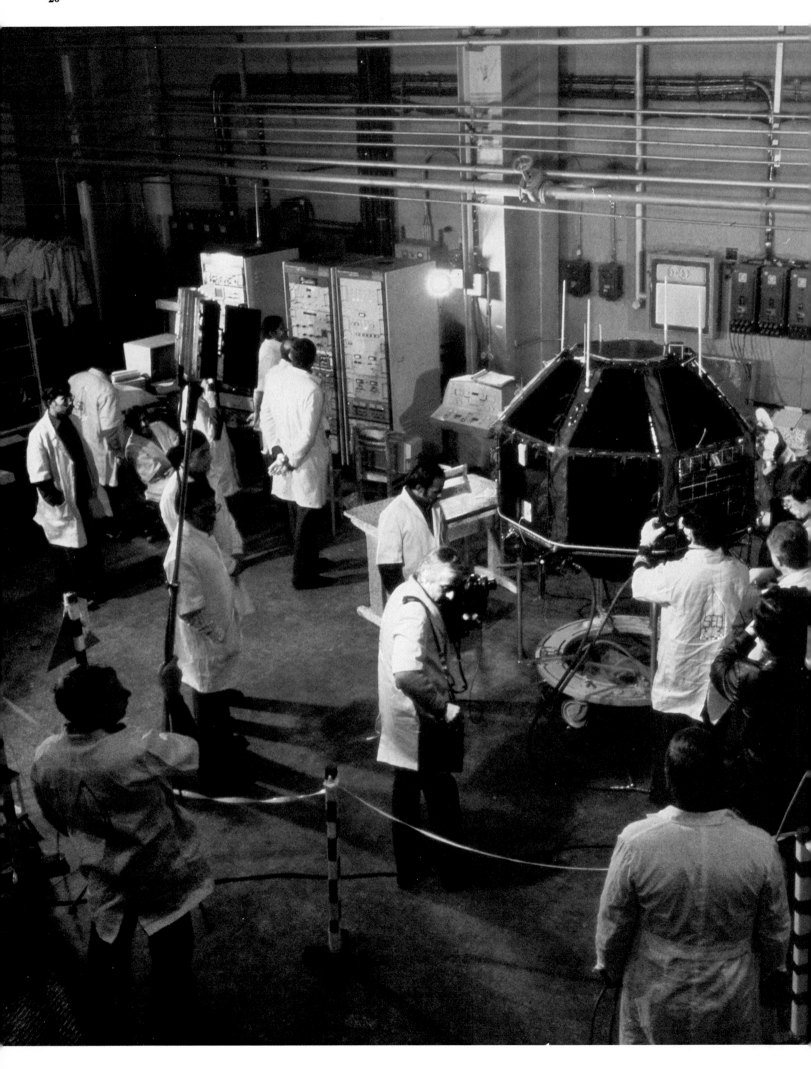

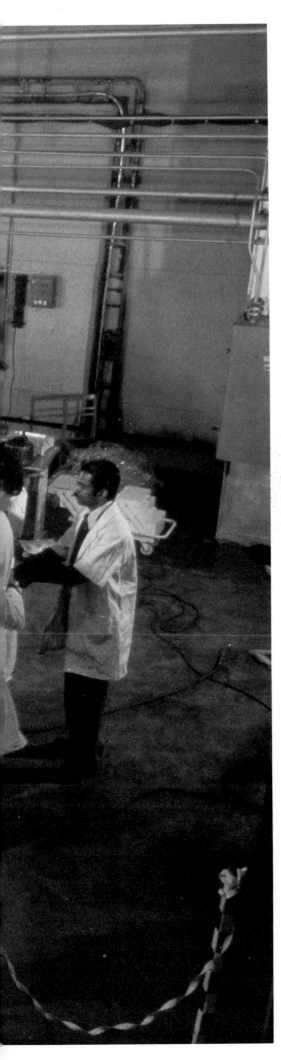

Ariabat

Launch Vehicle: C-1
Launch Site: Kapustin Yar
Launch Date: 19 April 1975
Total Weight: 360 kilograms
Apogee: 619 kilometers
Perigee: 563 kilometers
Inclination: 50.7 degrees
Period: 96.3 minutes

On 10 May 1972 the USSR signed an agreement with India to place an Indian satellite called Aryabhata in orbit. The launch took place on 19 April 1975 and was unusual in that some 50 Indian scientists and technicians were granted permission to observe the lift-off at Kapustin Yar. The Soviets handled the initial tracking efforts, but, once the orbit was well defined, joint Soviet-Indian tracking efforts were initiated. Ariabat's payload was designed to conduct geophysical experiments. Most of the spacecraft's components were manufactured in India, with the Soviets supplying solar cells and memory units. The instrument suite on Ariabat was used to conduct experiments concerning particle flows and radiation in the ionosphere, solar gamma and neutron radiation and x-ray astronomy. Power supply difficulties forced the termination of experiments after five days of operation.

Astron

Launch Vehicle: D-1e
Launch Site: Tyuratam
Launch Date: 23 March 1983
Apogee: 201,200 kilometers
Perigee: 1950 kilometers
Inclination: 51.09 degrees
Period: 98 hours

Astron is a modified Venera-type spacecraft designed for orbital astronomical observation. The vehicle was placed in a highly elliptical orbit in early 1983 so as to minimize disturbances that would adversely effect the operation of the spacecraft's x-ray and ultraviolet telescopes and to facilitate observations outside the interference of the Earth's Van Allen radiation belts. The x-ray telescope is a Roentgen design designated SKR-02 by the Soviets. It collects sources of x-ray radiation in the 2–25 kiloelectronvolt range, has a .2-square meter collecting area and ten spectral channels.

Spika, the ultraviolet telescope, is a joint French-Soviet project. Developed at the Cri-

mean Astrophysical Observatory, Spika is 4.2 meters long, weighs 400 kilograms and has a collecting surface area that Pravda claimed is a third greater than that of the US space telescope Copernicus, which operated from 1972 until 1981. Spika is a double reflecting telescope based on Richey-Chrétien optics. The main mirror has an 80-centimeter diameter, and the secondary mirror measures 26 centimeters in diameter. The telescope's entire focal length is eight meters.

Astron's collection activities are targeted primarily on the Taurus constellation. Taurus has two widely separated stellar concentrations and a crab-like nebula that contains a pulsar. In its first six months of operation Astron recorded radiation from 50 stars and 15 galaxies. The Soviets have also alluded to Astron's capabilities as evidence of their interest in the search for extraterrestrial intelligence.

Bhaskara

Bhaskara 1
Launch Vehicle: C-1
Launch Site: Kapustin Yar
Launch Date: 7 June 1979
Total Weight: 444 kilograms
Apogee: 557 kilometers
Perigee: 512 kilometers
Inclination: 50.7 degrees
Period: 95.2 minutes

Bhaskara 1 is an Indian earth resources satellite launched by the Soviet Union. Data from Bhaskara should help Indian agricultural planners predict shortages and surpluses so as to mitigate the consequences of marginal and bad harvests. In addition, earth resources data can help locate unexploited deposits of natural resources. Finally, information collected by Bhaskara's sensor package could have geodetic and cartographic applications.

Bhaskara 2
Launch Vehicle: C-1
Launch Site: Kapustin Yar
Launch Date: 20 November 1981
Apogee: 543 kilometers
Perigee: 521 kilometers
Inclination: 50.6 degrees
Period: 95.2 minutes

The only satellite launched from Kapustin Yar in 1981, Bhaskara 2's payload was placed on the southeasterly trajectory usually employed by Indian spacecraft launched by the Soviets. Bhaskara 2's mission is scientific. Its instrumentation includes a microwave radiometer and television equipment.

Below: A very well polished Elektron 1 mockup is shown on display in the Cosmos Pavilion at the Exhibition of Economic Achievements. This craft was designed to study the Van Allen radiation belts.

Ekran

Ekran, or Screen, is the Soviet name for a family of geosynchronous communications satellites that provide color and black-and-white television and radio programming to the remote northern and Siberian regions of the Soviet Union. The first Ekran was launched from Tyuratam on 26 October 1976 using a D-1e booster. From a near-Earth parking orbit, the satellite is placed in a geosynchronous orbit by a transfer rocket using an orbit that is tangential externally to the parking orbit at its perigee and internally to the final orbit at its apogee. The final apogee and perigee of the Ekran spacecraft are both 35,600 kilometers.

The total weight of the Ekran vehicle is 2700 kilograms. Its antenna consists of a 5.7×2.1-meter cophased helical array, which is deployed as 4×6 matrices mounted on four panels. The antenna array provides a gain of 28dB along the axis of the radiation pattern. The antenna is attached to a cylin-

drical main body that has a vernier rocket engine used to keep Ekran on station. Power is supplied by four solar cells mounted on booms extending from the main body at right angles. Each solar panel is composed of subassemblies deployed as 6x3 matrices. Ekran satellites remain in operation for two to three years, but more advanced models that can operate for at least five years may soon be placed in orbit. Two Ekrans are usually in operation at any given time. This constellation provides for transmission of one color television and two radio broadcast programs for 12 to 16 hours every 24 hours.

Elektron

The Elektron marked the first multiple-payload launches attempted by the Soviet Union. Two sets of Elektron doublets were orbited in 1964. The missions of these four spacecraft appear to have been primarily scientific, in that they were designed to map the Earth's high energy

Above: Elektron 1 was identical to Elektron 3 and, *at right:* Elektron 2 was identical to Elektron 4. Elektrons were paired 1–2, 3–4.

radiation (Van Allen) belts and provide synoptic readings. Some Western specialists have suggested that the satellites' highly elliptical orbits could have lent themselves to military-related experiments (*eg* nuclear explosion detection or early warning detection of ballistic missile launches).

Elektron 1 and 2

Launch Vehicle: A-1
Launch Site: Tyuratam
Launch Date: 30 January 1964
Total Weight, Elektron 1: 330 kilograms
Total Weight, Elektron 2: 445 kilograms
Apogee, Elektron 1: 7100 kilometers
Apogee, Elektron 2: 68,200 kilometers
Perigee, Elektron 1: 406 kilometers
Perigee, Elektron 2: 460 kilometers
Inclination, Elektron 1 and 2: 61 degrees
Period, Elektron 1: 169 minutes
Period, Elektron 2: 1360 minutes

СПУТНИК "ЭЛЕКТРОН-2"

СТАНЦИИ
«ЭЛЕКТРОН-I» И «ЭЛЕКТР

Elektron 3 and 4
Launch Vehicle: A-1
Launch Site: Tyuratam
Launch Date: 10 July 1964
Total Weight, Elektron 3: 330 kilograms
Total Weight, Elektron 4: 445 kilograms
Apogee, Elektron 3: 7040 kilometers
Apogee, Elektron 4: 66,235 kilometers
Perigee, Elektron 3: 405 kilometers
Perigee, Elektron 4: 459 kilometers
Inclination, Elektron 3 and 4: 60.86 degrees
Period, Elektron 3: 168 minutes
Period, Elektron 4: 1314 minutes

Elektron 1 and 3 were cylindrical satellites powered by six solar panels deployed at differing angles to the axis of the spacecraft. Several antennas protruded from either end. Elektron 2 and 4 were nipple-shaped, with power provided by solar panels imbedded in the main body of the satellites. Antennas were positioned at both ends of the spacecraft, and a magnetometer boom was positioned at the tapered end of the satellites. Elektron 1 and 3 were separated from the final stage of the A-1 booster prior to thrust termination by a small solid-propellant rocket motor attached to the base of the spacecraft. Elektron 3 and 4 continued on with the final stage until they reached their more eccentric orbits.

Gorizont

The Gorizont, or Horizon, family of Soviet geosynchronous communications satellites made its debut on 19 December 1978. Tass declared that the USSR planned 'to use Gorizont satellites for relaying Soviet TV transmissions of the Twenty-second Summer Olympics in 1980.' The first Gorizont, however, was at least a partial failure, in that it never achieved a truly geostationary orbit. Beginning with the second Gorizont, which was launched on 5 July 1979, these satellites were placed in true geosynchronous orbits with both apogee and perigee at 36,550 kilometers.

Gorizont satellites weigh 2800 kilograms and have a main body and annular thermoregulator system similar to that of the Ekran family of satellites. Solar panels located on arrays connected to the main body provide 2.3 square meters of collecting surface. Other panels are located in cells placed along the circumference of the upper portion of the main body of the vehicle. Three infrared and one visible light horizon sensors provide the necessary data to maintain three-axis body stabilization of the spacecraft. Global coverage horn antennas and spot-beam receiving and transmitting antennas are mounted on the Gorizont vehicle. A 4-GHz downlink is supplied by both a large circular dish antenna and the larger of three 'orange peel' antennas. A 6-GHz uplink is provided by the two smaller 'orange peel' shaped antennas. A 40-W repeater and five 15-W repeaters are mounted on Gorizont.

Interkosmos

The Soviets have sought from the beginning of their space program to maximize the prestige value associated with exploration of the cosmos and development of space manufacturing applications. Nor has the propaganda value of linking socialist economic organization with success in space escaped the Politburo. Since socialism extends beyond the frontiers of the USSR, the Soviet leadership has made a concerted effort to involve allies and client states in at least the civilian and scientific aspects of their space program. In 1967 the Soviets founded the Interkosmos program with the aim of engendering cooperative space research efforts among the members of the Eastern bloc. The original members were Bulgaria, Cuba, Czechoslovakia, East Ger-

Below: The instrument hub bristles in front of the solar panel of this Interkosmos satellite. *Right:* The Czech-Soviet-East German Interkosmos 1 being prepared for launch.

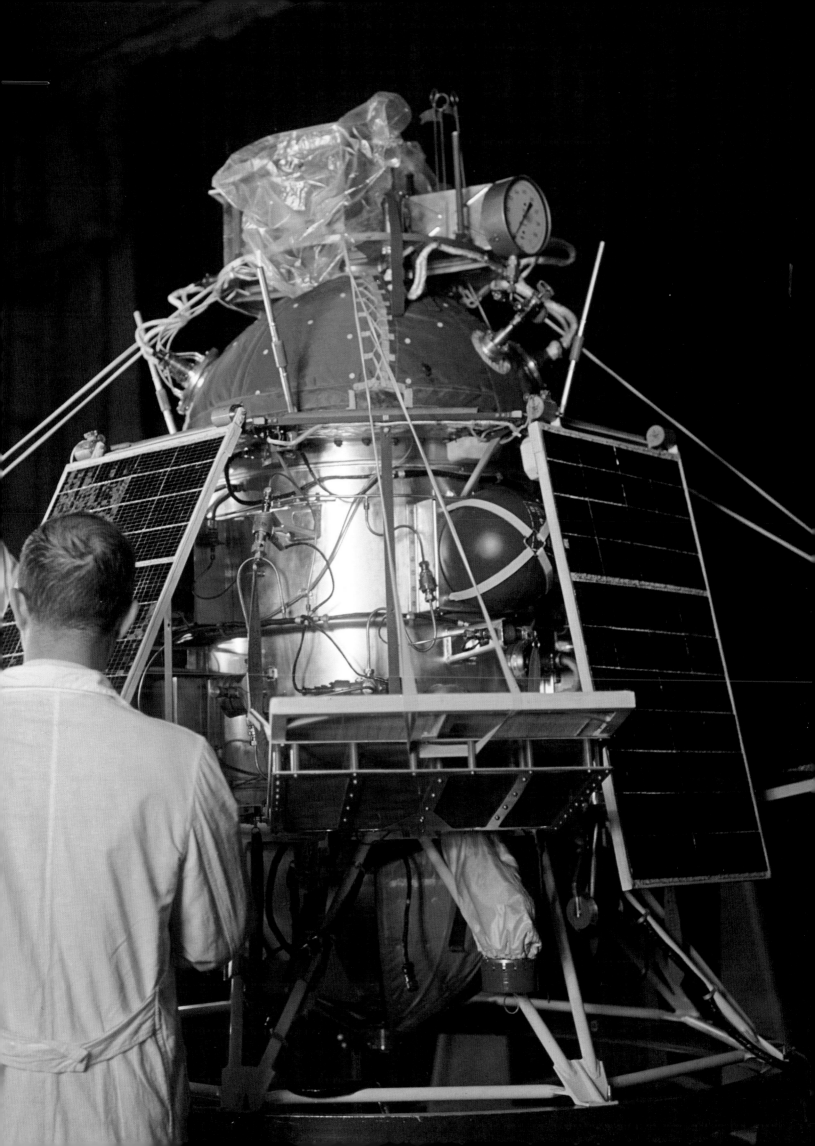

Borealis. Although the payload was multinational, none of the Warsaw Pact countries was allowed to send observers to Plesetsk for the launch.

Interkosmos 1

Launch Vehicle: B-1
Launch Site: Kapustin Yar
Launch Date: 14 October 1969
Decay Date: 2 January 1970
Total Weight: 320 kilograms
Apogee: 640 kilometers
Perigee: 260 kilometers
Inclination: 48.4 degrees
Period: 93.3 minutes

The Interkosmos 1 payload contained equipment from three Warsaw Pact nations: Czechoslovakia, East Germany and the USSR. The satellite contained experiments designed to measure the effects of solar ultraviolet and x-ray radiation upon the composition of the upper atmosphere. Scientists from East Germany, Hungary, Mongolia, Poland, Rumania and the USSR analyzed the resulting data.

Interkosmos 2

Launch Vehicle: B-1
Launch Site: Kapustin Yar
Launch Date: 25 December 1969
Decay Date: 7 June 1970
Total Weight: 320 kilograms
Apogee: 1200 kilometers
Perigee: 206 kilometers
Inclination: 48.4 degrees
Period: 98.5 minutes

The Interkosmos 2 payload was a product of Bulgarian, Czechoslovakian, East German and Soviet cooperation. The mission measured the density of positive ions and electrons in the upper atmosphere, the mean electron concentration between the spacecraft and the ground stations monitoring the flight and the electronic temperature in close proximity to the satellite. The resulting data was shared with the rest of the Warsaw Pact nations and Cuba. Two tracking stations in Poland were employed, in addition to seven located in the Soviet Union.

many, Hungary, Mongolia, Poland, Rumania and the USSR. The Socialist Republic of Vietnam joined in 1979.

The Interkosmos program is not the only cooperative space effort that has been undertaken by the Soviet Union. The Soviets have also cooperated with France, India, Sweden and the United States in various space ventures. The efforts that included non-Soviet bloc countries and those that involved socialist nations under other Soviet spacecraft programs (eg, Soyuz) are described elsewhere in this volume. Kosmos launches that do not fit into the above categories are considered in this section.

Kosmos 261

Launch Vehicle: B-1
Launch site: Plesetsk
Launch Date: 19 December 1968
Decay Date: 12 February 1969
Total Weight: 347 kilograms
Apogee: 670 kilometers
Perigee: 217 kilometers
Inclination: 71 degrees
Period: 93.1 minutes

Kosmos 261 contained a payload developed in a cooperative effort by seven Soviet bloc countries: Bulgaria, Czechoslovakia, East Germany, Hungary, Poland, Rumania and the USSR. The geophysical experiments conducted were designed to examine the upper atmosphere, the Northern Lights, the behavior of electrons and protons at the edge of space, electronic aspects of super thermal energy and atmospheric density variations associated with the Aurora

Kosmos 348

Launch Vehicle: B-1
Launch Site: Plesetsk
Launch Date: 13 June 1970
Total Weight: 357 kilograms
Apogee: 680 kilometers
Perigee: 212 kilometers
Inclination: 71 degrees
Period: 93 minutes

Designed to study corpuscular streams in the Earth's ionosphere, the payload for Kosmos 348 was the product of a cooperative effort involving Bulgaria, Czechoslovakia, East Germany, Hungary, Poland, Rumania and the Soviet Union. The Plesetsk launch was barred to non-Soviet observers. (Plesetsk was not opened to non-Soviet personnel until 1972.)

Interkosmos 3

Launch Vehicle: B-1
Launch Site: Kapustin Yar
Launch Date: 7 August 1970
Decay Date: 6 December 1970
Total Weight: 340 kilograms
Apogee: 1320 kilometers
Perigee: 207 kilometers
Inclination: 49 degrees
Period: 99.8 minutes

A payload of Czechoslovak and Soviet instruments was orbited in this spacecraft for the purpose of studying cosmic rays and charged particles. Specific experiments included measuring interactions between solar activity and the Van Allen belts and the spectrum of low frequency electromagnetic oscillations in the upper atmosphere.

Interkosmos 4

Launch Vehicle: B-1
Launch Site: Kapustin Yar
Launch Date: 14 October 1970
Decay Date: 7 January 1971
Total Weight: 320 kilograms
Apogee: 668 kilometers
Perigee: 263 kilometers
Inclination: 48.5 degrees
Period: 93.6 minutes

Interkosmos 4 was essentially a clone of the first Interkosmos flight. The instruments supplied by Czechoslovakia, East Germany and the USSR were more sensitive, however, and capable of measuring solar ultraviolet and x-ray radiation with greater specificity.

Interkosmos 5

Launch Vehicle: B-1
Launch Site: Kapustin Yar
Launch Date: 2 December 1971
Decay Date: 7 April 1972
Total Weight: 340 kilograms
Apogee: 1200 kilometers
Perigee: 205 kilometers
Inclination: 48.4 degrees
Period: 98.5 minutes

Interkosmos 5 was a follow-on to Interkosmos 3, with emphasis upon cosmic

Left: Interkosmos 2 and a B-1 on the pad. At Right: The Interkosmos 4 is here mounted to its B-1 launch vehicle. The orange object in the foreground is a protective capsule, to be jettisoned previous to satellite orbit.

rays and the flux created by charged particles. Czech and Soviet equipment comprised the payload, while ground stations located in Czechoslovakia, East Germany and the USSR supported the flight with synoptic readings.

Interkosmos 6
Launch Vehicle: A-2
Launch Site: Tyuratam
Launch Date: 7 April 1972
Recovery Date: 11 April 1972
Total Weight: 5700 kilograms
Apogee: 256 kilometers
Perigee: 203 kilometers
Inclination: 51.8 degrees
Period: 89 minutes

Interkosmos 6 was the only spacecraft in the series that was recovered. The payload was designed to conduct a study of high-energy cosmic rays. Specifications for a photoemulsion unit and ionization calorimeter were developed in Czechoslovakia, Hungary, Mongolia, Poland, Rumania and the USSR. These specifications were turned into hardware in the Soviet Union. A joint Czech, Hungarian and Soviet effort produced a meteorite experiment that was also flown on the spacecraft. Recovery allowed a much more detailed analysis of the data and instruments than would otherwise have been possible.

Interkosmos 7
Launch Vehicle: B-1
Launch Site: Kapustin Yar
Launch Date: 30 June 1972
Decay Date: 5 October 1972
Total Weight: 375 kilograms
Apogee: 568 kilometers
Perigee: 267 kilometers
Inclination: 48.4 degrees
Period: 92.6 minutes

A continuation of the Interkosmos 1 and 4 missions, this satellite also carried a payload (a Czech, East German and Soviet product) designed to measure short wave and hard x-ray solar radiation. The vehicle was also able to collect data on numerous solar flares that could not be observed by groundbased equipment.

Interkosmos 8
Launch Vehicle: B-1
Launch Site: Plesetsk
Launch Date: 1 December 1972
Decay Date: 2 March 1973
Total Weight: 340 kilograms
Apogee: 679 kilometers
Perigee: 214 kilometers
Inclination: 71 degrees
Period: 93.2 minutes

The first of the Interkosmos series to be launched from Plesetsk, Interkosmos 8 also marked the lifting of restrictions on Warsaw Pact mission specialists that prevented them from attending the launch of cooperative payloads from the northernmost of the Soviet Kosmodromes. Bulgaria contributed an ion trap and a Langmuir probe. Czech specialists designed and built a high frequency probe. East Germany supplied a Mayak transmitter and recorder and the Soviets contributed an ionospheric gas discharge counter, among other instruments, to the payload. The mission collected data on high energy protons, ions and electrons in the upper reaches of the Earth's atmosphere.

Interkosmos 9 (Kopernik 500)
Launch Vehicle: B-1
Launch Site: Kapustin Yar
Launch Date: 19 April 1973
Decay Date: 19 October 1973
Total Weight: 340 kilograms
Apogee: 1551 kilometers
Perigee: 202 kilometers
Inclination: 48.5 degrees
Period: 102.2 minutes

Launched on the 500th birthday of the Polish astronomer Copernicus, Interkosmos 9 carried equipment from his homeland developed in cooperation with the Soviet Union. During its five months of life, the satellite payload performed measurements of the ionosphere and solar radiation. Czech scientists contributed the telemetry package, and data was recorded at ground sites in Czechoslovakia and the USSR.

Interkosmos 10
Launch Vehicle: C-1
Launch Site: Plesetsk
Launch Date: 30 October 1973
Decay Date: 1 July 1977
Total Weight: 340 kilograms
Apogee: 1477 kilometers
Perigee: 265 kilometers
Inclination: 74 degrees
Period: 102 minutes

The introduction with the launch of Interkosmos 10, of the more powerful C-1 booster (see Appendix II for details) provided the Interkosmos program greater flexibility with respect to payload and orbital parameters. The instrumentation for this satellite was of Czech, East German and Soviet design. The mission was to conduct measurements of the concentration and tem-

At right: Interkosmos 4. The Interkosmos program of international orbital projects fosters an increasing openness in the USSR to international manned space projects. **Note** instruments and sensors on craft's nose.

perature of electrons in the ionosphere, determine variations in the Earth's magnetic field and to gather data on the low frequency electric fluctuations of plasma.

Interkosmos 11

Launch Vehicle: C-1
Launch Site: Kapustin Yar
Launch Date: 17 May 1974
Total Weight: 350 kilograms
Apogee: 526 kilometers
Perigee: 484 kilometers
Inclination: 50.7 degrees
Period: 94.5 minutes

The first C-1 launch of the series from Kapustin Yar, Interkosmos 11 continued the program of study concerning solar ultra violet and x-ray radiation undertaken by Interkosmos 1, 4 and 7. In addition, the payload measured magnetosphere interactions. The instrumentation was a product of Czechoslovak, East German and Soviet cooperation.

Interkosmos 12

Launch Vehicle: C-1
Launch Site: Plesetsk
Launch Date: 31 October 1974
Decay Date: 11 July 1975
Total Weight: 350 kilograms
Apogee: 708 kilometers
Perigee: 264 kilometers
Inclination: 74.1 degrees
Period: 94.1 minutes

An all-Warsaw Pact effort, Interkosmos 12 contained Bulgarian and Soviet equipment for measuring the temperature of positive ions and electrons; mass spectrometers of Czech and Soviet design; a Romanian calibration instrument for the mass spectrometers; a Mayak radio transmitter made in Czechoslovakia; an instrument built by Czech, East German and Soviet specialists for determining electron concentration densities; a memory unit from East Germany; and a Hungarian, Czech and Soviet device for analyzing micrometeorites. This multinational geophysical payload was designed to conduct measurements of the atmosphere and ionosphere and to study the flow of micrometeorites in near-Earth orbit.

Interkosmos 13

Launch Vehicle: C-1
Launch Site: Plesetsk
Launch Date: 27 March 1975
Decay Date: 2 September 1980
Total Weight: 350 kilograms
Apogee: 1714 kilometers
Perigee: 296 kilometers
Inclination: 83 degrees
Period: 104.9 minutes

This satellite was a joint Czech-Soviet mission to observe the dynamics of the polar ionosphere and the magnetosphere.

Interkosmos 14

Launch Vehicle: C-1
Launch Site: Plesetsk
Launch Date: 11 December 1975
Total Weight: 372 kilograms
Apogee: 1707 kilometers
Perigee: 345 kilometers
Inclination: 74 degrees
Period: 105.3 minutes

A four-nation payload from Bulgaria, Czechoslovakia, Hungary and the USSR, Interkosmos 14's mission closely resembled that of Interkosmos 12, with an emphasis upon the level of micrometeorite activity and electron flux in the magnetosphere.

Interkosmos 15

Launch Vehicle: C-1
Launch Site: Plesetsk
Launch Date: 19 June 1976
Decay Date: 18 November 1979
Total Weight: 422 kilograms
Apogee: 521 kilometers
Perigee: 487 kilometers
Inclination: 74 degrees
Period: 94.6 minutes

The value of the more capable C-1 booster was revealed in this launch, which saw the inauguration of a new generation of geophysical payloads. The automatic universal orbital station, dubbed AUOS, was a major improvement over previous Interkosmos vehicles, allowing uplink commands to be generated at any point in the satellite's orbit (previous Interkosmos missions could only receive commands within line-of-sight of controlling ground stations). In addition, the AUOS provided for larger payloads. The first AUOS payload was a product of Warsaw Pact cooperation. It contained a new telemetry system (known by the acronym YeTMS) that was developed and manufactured by Czechoslovakia, East Germany, Hungary, Poland and the USSR. The YeTMS featured onboard processing of data, which was then digitally transmitted to reception sites in Bulgaria, Cuba, Czechoslovakia, East Germany, Hungary and the Soviet Union.

Interkosmos 16

Launch Vehicle: C-1
Launch Site: Kapustin Yar
Launch Date: 27 July 1976
Decay Date: 10 July 1979
Total Weight: 475 kilograms
Apogee: 523 kilometers
Perigee: 465 kilometers

Inclination: 50.6 degrees
Period: 94.4 minutes

This satellite was the last in the series of Interkosmos and Vertikal missions concerned primarily with the effects of solar x-rays and ultraviolet radiation upon the upper reaches of the Earth's atmosphere. For the first time in the Interkosmos program, however, the cooperative scientific efforts of the Warsaw Pact nations were enhanced by the participation of a Western nation. Sweden supplied a spectropolarimeter for studying solar flare activity. The Soviet and Eastern European contributions to the Interkosmos 16 payload are listed below:

· a Czech multichannel photometer for observation of solar flares
· an East German photometer for measuring densities of molecular oxygen in the upper atmosphere
· a spectroheliograph of Soviet make for studying solar x-ray flares

Satellite experiments were complemented by coincident ground based solar observations taken from installations in Bulgaria, Czechoslovakia, East Germany, Hungary and the USSR.

Interkosmos 17

Launch Vehicle: C-1
Launch Site: Plesetsk
Launch Date: 24 September 1977
Decay Date: 8 November 1979
Total Weight: 550 kilograms
Apogee: 519 kilometers
Perigee: 468 kilometers
Inclination: 83 degrees
Period: 94.4 minutes

This mission marked the first operational use of the AUOS payload vehicle and the YeTMS digital telemetry system. Interkosmos 17 was a follow-on in the research program associated with the third, fifth and thirteenth launches of the program that were designed to observe the interactions of phenomena created by solar activity with the Earth's upper atmosphere. The geophysical instrument package was a cooperative Warsaw Pact effort. Payload experiments included measurements of micrometeorite intensity and readings concerning charged and neutral particle densities. Specific equipment included the following:

· two Bulgarian, Czech and Soviet dosimeters
· an instrument from Czechoslovakia that measured isotopes of solar streams
· equipment made by Czechoslovakia and the Soviet Union for determining the tem-

perature of electrons in the ionosphere
· neutron-measuring instruments from Czechoslovakia and the Soviet Union
· a Czech and Soviet-made electronic analyzer
· a Romanian and Soviet spectrometer
· Soviet differential proton and electron detectors studying weak cosmic radiation

Interkosmos 18
Launch Vehicle: C-1
Launch Site: Plesetsk
Launch Date: 24 October 1978
Total Weight: 550 kilograms
Apogee: 786 kilometers
Perigee: 407 kilometers
Inclination: 83 degrees
Period: 96.4 minutes

Above: **The Interkosmos emblem, with the Soviet star rising into orbit to encircle the Earth.**

Interkosmos 18 was actually a dual-satellite launch comprising the main payload and MAGION, which was the first Czechoslovak satellite to reach near-Earth orbit successfully. With the exception of Soviet systems for regulating power and heat, MAGION was developed and built exclusively by Czech scientists and technicians. MAGION weighed 15 kilograms, had an apogee of 762 kilometers, a perigee of 404 kilometers, an inclination of 83 degrees and an orbital period of 96.3 minutes. It was separated from the mother ship 21 days into the mission. Its very similar orbital parameters facilitated temporal and spacial differentiation with re-

spect to measurements taken by the two vehicles. When MAGION came within line-of-sight of Czech ground stations, its usual data transmissions were replaced by musical tones.

Both satellites conducted experiments aimed at defining the electromagnetic relationships between the ionosphere and the magnetosphere. In addition, experiments were conducted concerning low-frequency radio waves in the circumterrestrial plasma. Equipment contributions are listed below by country.

· Czechoslovakia and the USSR: a mass spectrometer, a low-frequency analyzer for electromagnetic fields and instruments for determining electron temperatures

· East Germany: Langmuir probes for measuring the temperature of electrons in plasma and a flat ion trap to measure ion temperatures and density
· Romania and the USSR: a SG-R magnetometer
· USSR: a charged-particle electrostatic analyzer and a plasma electric parameter analyzer

Interkosmos 19

Launch Vehicle: C-1
Launch Site: Plesetsk
Launch Date: 27 February 1979
Total Weight: 550 kilograms
Apogee: 996 kilometers
Perigee: 502 kilometers
Inclination: 74 degrees
Period: 99.8 minutes

Interkosmos 19 joined its immediate predecessor, MAGION, two American geophysical satellites and Swedish, French, Australian and Soviet balloon- and sounding-rocket-carried instrument packages in a major cooperative program to study the Earth's magnetosphere. In addition to the participating nations, Japan joined in the analysis of the resulting data. Individual Warsaw Pact nation contributions to the Interkosmos 19 payload are listed below.

· Bulgaria: an optical spectrometer and instruments for determining electron temperatures and densities in magnetosphere
· Czechoslovakia: high and low frequency probes for determing distribution and temperature of thermal electrons
· Poland: a radiospectrometer
· USSR: an ionospheric station with instrumentation for collecting data on photoelectric pulses and electron showers, a Mayak transmitter and a device for analyzing plasma flows

Interkosmos 20

Launch Vehicle: C-1
Launch Date: Plesetsk

Launch Date: 1 November 1979
Total Weight: 80 kilograms
Apogee: 523 kilometers
Perigee: 467 kilometers
Inclination: 74 degrees
Period: 94.4 minutes

The first applications mission in the Interkosmos program, Interkosmos 20 conducted oceanographic research functions while serving as a data relay between various buoys and platforms deployed on the ocean surface and mission control at Tarusa in the USSR. The flight was a cooperative Czech, East German, Hungarian and Soviet undertaking that focused on locating areas of significant maritime bio-activity and on measuring oceanic surface temperatures from space.

Interkosmos 21

Launch Vehicle: C-1
Launch Site: Plesetsk
Launch Date: 6 February 1981
Apogee: 520 kilometers
Perigee: 475 kilometers
Inclination: 74 degrees
Period: 94.5 minutes

An oceanographic mission like that of its immediate predecessor, Interkosmos 21 continued efforts to locate high concentrations of marine life and maritime pollution and to collect ocean surface temperature data. In addition, this flight conducted experiments concerning land water, water ice and snow-clear area boundaries. Readings were taken in different spectral bands of the atmosphere in order to determine its optical thickness. Czechoslovakia, East Germany, Hungary, Romania and the USSR cooperated on development and construction of the payload instruments. Specific equipment comprised systems designed to collect and transmit data received from ground stations located in Baku, Budapest, East Berlin, the Indian Ocean, Moscow, Sebastopol and Vladivostok: a multichannel spectrometer, a bipolar

radiometer and a three-component magnetometer.

Interkosmos 22—Bulgaria 1300

Launch Vehicle: A-1
Launch Site: Plesetsk
Launch Date: 7 August 1981
Apogee: 893 kilometers
Perigee: 799 kilometers
Inclination: 81.2 degrees
Period: 101.8 minutes

The second Interkosmos launch to use the large A-1 booster, this mission was undertaken in conjunction with celebrations of the settlement of Bulgaria by Bulgar tribesmen in 681 BC. The joint Soviet-Bulgarian instrumentation package conducted familiar Interkosmos geophysical experiments concerning ionospheric and magnetospheric research. Soviet equipment conducted Earth resources observations, while Bulgarian instruments took readings of the Earth's upper atmosphere in conjunction with measurements provided by a Bulgarian instrument package on a Meteor-Priroda vehicle launched a month earlier (see the Meteor-Priroda heading in this chapter) and fourteen ground-based laser installations. Bulgarian equipment included a system for determining the temperature, energy distribution and ion mass of electrons; instruments for measuring electric fields; an electrophotometer that detected weak light emissions; an instrument for measuring ultraviolet radiation in the upper atmosphere; and an angular laser reflector for geodetic measurements.

Korabl Sputnik

The five spacecraft in this series were precursors of the manned Vostok program. The Soviets sought to test the ability to recover satellites from earth orbit, an essential prerequisite to manned flight. The USSR also sought to determine what stresses its cosmonauts would be subject to upon reentry. These tests were accom-

Recoverable Kosmos Launches

	1962	1963	1964	1965	1966	1967	1968	1969	1970	1971	1972	1973
Tyuratam Launch	5	7	12	17	15	8	14	13	13	13	11	8
Plesetsk Launch	—	—	—	—	6	14	15	18	16	15	18	27
Total Recoverable	5	7	12	17	21	22	29	31	29	31	29	35
Total Kosmos	12	12	23	36	36	61	64	55	65	67	58	64
Annual % Recoverable	42	58	52	47	58	36	45	56	45	42	50	55
Cumulative Recoverable %	41	50	51	49	52	47	46	48	48	47	47	48

plished using dogs and dummies that were observed via realtime television links and biomedical telemetry. By the spring of 1961 the Soviets considered the Korabl Sputnik to be man-rated, and Yuri Gagarin embarked on his historic flight (see the entry under Vostok).

Korabl Sputnik 1
Launch Vehicle: A-1
Launch Site: Tyuratam
Launch Date: 15 May 1960
Decay Date: 9 September 1962
Total Weight: 4540 kilograms
Apogee: 369 kilometers (subsequently 657 kilometers)
Perigee: 312 kilometers (subsequently 290 kilometers)
Inclination: 65 degrees
Period: 91.2 minutes (subsequently 94.3 minutes)

This flight was the first application of the A-1 booster to near-Earth orbit. (It had previously been employed for direct Lunar ascent missions: See Appendix II for details) The payload consisted of a 2500-kilogram reentry vehicle containing life-support equipment and a 1477-kilogram instrumentation package. The reentry capsule returned telemetry, prerecorded tapes of a Russian choral group (so as to avoid Western accusations that the flight was actually a manned mission that failed) and television coverage of the capsule interior for observing the effects of reentry upon the dummy. The first Korabl Sputnik was at least a partial failure, in that an incorrect orientation during the recovery attempt resulted in a high orbit that did not completely decay for five years. The service module reentered on 5 September 1962, with some fragments reaching the ground near Manitowoc, Wisconsin.

Korabl Sputnik 2
Launch Vehicle: A-1
Launch Site: Tyuratam
Launch Date: 19 August 1960

Recovery Date: 20 August 1960
Total Weight: 4600 kilograms
Apogee: 339 kilometers
Perigee: 306 kilometers
Inclination: 65 degrees
Period: 90.7 minutes

Korabl Sputnik 2 was the first successfully recovered satellite in history. Its two passengers, the dogs Strelka and Belka, became national heroes. In addition to the dogs, the spacecraft carried some 40 mice. The capsule was deorbited so as to land in the Soviet Union, a recovery pattern that obtains for most recoverable Soviet spacecraft to this day.

Korabl Sputnik 3
Launch Vehicle: A-1
Launch Site: Tyuratam
Launch Date: 1 December 1960
Decay Date: 2 December 1960
Total Weight: 6483 kilograms
Apogee: 265 kilometers
Perigee: 188 kilometers
Inclination: 65 degrees
Period: 88.6 minutes

The passengers of Korabl Sputnik 3, the dogs Pchelka and Mushka, were not as fortunate as their predecessors. The recovery attempt failed because the reentry angle was too steep, and the capsule burned up. Thus Pchelka and Mushka joined Layka (see the Sputnik 2 entry) as the earliest canine casualties of orbital flight. The capsule was not ready for Yuri Gagarin.

Korabl Sputnik 4
Launch Vehicle: A-1
Launch Site: Tyuratam
Launch Date: 9 March 1961
Recovery Date: 9 March 1961
Total Weight: 4700 kilograms
Apogee: 249 kilometers
Perigee: 184 kilometers
Inclination: 64.9 degrees
Period: 88.5 minutes

Soviet mission controllers successfully recovered this spacecraft following a single orbit. The reentry vehicle contained the dog Chernushka and a dummy cosmonaut.

Korabl Sputnik 5
Launch Vehicle: A-1
Launch Site: Tyuratam
Launch Date: 25 March 1961
Recovery Date: 25 March 1961
Total Weight: 4695 kilograms
Apogee: 247 kilometers
Perigee: 178 kilometers
Inclination: 64.9 degrees
Period: 88.4 minutes

In hindsight it is evident that the Soviet pattern of recovery following a single orbit, which was continued with Korabl Sputnik 5, was a preparation for Gagarin's single-orbit flight that was to follow in less than a month. This flight carried a dummy and the dog Zvezdochka.

Kosmos (Military)

The Soviets launched the first of their Kosmos series of spacecraft on 16 March 1962. While the announced purpose of the Kosmos program was 'to continue the study of outer space,' the term itself has been employed primarily to mask the actual mission of the majority of Soviet space launches for the last two and a half decades. Kosmos launches are routinely accompanied by bland announcements concerning date of launch and orbital parameters for the vehicle. This is generally the only information that the Soviets supply concerning the bulk of their space launches.

The reason for such reticence has to do with the sensitive military nature of the overwhelming majority of missions conducted under the Kosmos designator. Around 70 percent of all Soviet space systems have a purely military mission. Perhaps an additional 15 percent of the Soviet space program can be characterized as serving a dual civil-

Recoverable Kosmos Launches (Continued)

	1974	1975	1976	1977	1978	1979	1980	1981	1982	1983	1984	1985
Tyuratam Launch	9	12	13	10	8	2	3	17	15	10	8	10
Plesetsk Launch	19	21	21	23	27	34	32	20	20	27	28	24
Total Recoverable	28	33	34	33	35	36	35	37	35	37	36	34
Total Kosmos	60	64	79	79	67	64	67	72	82	79	79	82
Annual % Recoverable	47	52	43	42	52	56	52	51	43	47	46	41
Cumulative Recoverable %	48	48	48	47	48	48	48	52	51	51	51	50

Source: G.E. Perry, "Identification of military components within the Soviet space programme" (Kettering, U.K.: The Kettering Group, 1982); Nicholas L. Johnson, *The Soviet Year in Space*, years 1981–1985 (Colorado Springs, CO: Teledyne Brown Engineering, 1982–1986).

percent of all Soviet space systems have a purely military mission. Perhaps an additional 15 percent of the Soviet space program can be characterized as serving a dual civilian/military purpose. These figures alone tend to explain the large number of Kosmos launches conducted each year with only the most cursory of explanations.

Soviet military space missions can be divided into nine categories: photographic intelligence (PHOTINT), electronic intelligence (ELINT), ocean reconnaissance, early warning, communications, navigation/geodesy, minor military, antisatellite (ASAT) and fractional orbital bombardment systems (FOBS). The remainder of this section catalogues and describes the Soviet spacecraft employed to carry out these missions. The next section will examine those Kosmos launches that appear to have scientific and space applications missions.

Photographic Reconnaissance Satellites

On 26 April 1962 Kosmos 4 was launched from Tyuratam atop an A1 booster into a 330×298-kilometer orbit. The 4600-kilogram vehicle had an inclination of 65 degrees and an orbital period of 90.6 minutes. On 29 April the payload was deorbited and recovered, thus

ending the first successful Soviet photo-reconnaissance mission. Roughly half of all Kosmos launches since Kosmos 4 have been recovered from orbit. PHOTINT missions comprise the largest single mission category in the Soviet space program. To date, the Soviets have developed five generations of photo-reconnaissance satellites, which are described below. Each new generation has expanded Soviet capability to monitor the military forces of its adversaries, track the course of international crises and conflicts and verify compliance with arms control agreements by the acceding parties. Table 2 catalogues the growth of Soviet PHOTINT capabilities in terms of days per year in which the USSR had a reconnaissance satellite in orbit.

First Generation Photographic Reconnaissance Satellites

As mentioned above, the Soviets launched their first photo-reconnaissance mission in 1962. The initial generation of this type of spacecraft was based on a modified Vostok spacecraft of the kind that carried Yuri Gagarin on his historic flight. The total weight of the vehicle was approximately 4600 kilograms. The camera apparatus was housed in the spherical reentry module

The national flags of project participants encircle the apex of Kosmos 782, *below*. *Right*: Scientists prepare to disembark monkeys Abrek and Bion from Kosmos 1514.

which weighed some 2300 kilograms. The reentry module was deorbited by a retrorocket system which was shaped like a double cone and weighed some 2200 kilograms. Most flights during 1962 were devoted to research and development, with an average length of four to six days. The early flights of this generation employed the A1 booster which had launched the original Vostoks.

The first operational photo-reconnaissance mission was Kosmos 12, which was launched on 22 December 1962 and recovered eight days later. The eight-day lifetime was standard for the first generation PHOTINT systems (although some 10-day missions were accomplished). These spacecraft were not maneuverable, and it can be assumed that their photographic resolution capability was poor compared to today's collection systems. Initially, the first generation vehicles employed the standard Vostok inclination of 65 degrees, however, in 1964 missions began using an inclination of 51.2 degrees.

In late 1963 the Soviet reconnaissance program began using the A-2 booster, which has been used to launch all succeeding generations of PHOTINT systems. Kosmos 112 inaugurated the new Plesetsk Kosmodrome on 17 March 1966, and within a year the northern launch site had surpassed Tyuratam in number of annual launches—a situation which obtains to this day. The usual inclination for first generation launches from Plesetsk was 64.6 degrees.

First generation payloads were recovered using standard Vostok reentry techniques. Orientation for retrofire was determined by an optical horizon-sensing package. After the retrorockets began the deorbiting process, the recovery capsule was separated from the instrument/retrorocket package, which continued to transmit telemetry until it burned up in the atmosphere. Meanwhile the recovery module continued its ballistic reentry. At a height of eight or nine kilometers the capsule's parachute deployed and a recovery beacon was activated. Measurements of the length and strength of the recovery signal (Morse code TK groups transmitted on 19.995 MHz and 20.005 MHz) indicated a seven-minute descent interval. Differences in the recovery beacons

used by Soviet photo-reconnaissance satellites have facilitated classification of different mission types by Western experts. The final first generation probably had low-resolution and medium-resolution variants. The last first generation photo-reconnaissance mission was probably Kosmos 153, which was launched from Plesetsk on 4 April 1967.

Second Generation Photographic Reconnaissance Satellites

Kosmos 22 marked the debut of the second generation of Soviet reconnaissance satellites. By using the A-2 booster, the USSR could put a 'heavy' Vostok spacecraft into orbit, with a total weight of some 5500 to 6000 kilograms. Kosmos 22 was a high-resolution system which had many of the features of the first generation vehicles, including an eight-day operational life, no maneuvering capability and a similar recovery mode. The final second generation high-resolution mission was Kosmos 355, launched from Plesetsk on 7 August 1970.

A low-resolution second generation variant was first placed in orbit on 8 June 1966 (Kosmos 120). Other than the type of photography produced by its camera system (area- as opposed to point-target collection), this system was little different from its predecessor, and both of the second generation variants were essentially large versions of the first generation vehicles. The final launch of a second generation low-resolution mission occurred at Plesetsk on 12 May 1970 (Kosmos 344).

The final second generation variant proved to be a major design innovation. This spacecraft was termed 'extended-duration' because its standard lifetime exceeded the high- and low-resolution variants by 50 per cent. Its recovery procedure also differed from those employed in the past. Following recovery, Western tracking facilities usually observed debris left in orbit. Eventually it was determined that the Soviets had further modified the Vostok vehicle to accommodate a biological experimentation package, which was mounted above the reentry module. Thus this photo-reconnaissance satellite had a dual military/civilian role. The biolog-

ical package could not have survived reentry, so it was jettisoned at the time of retrofire. The first extended duration flight occurred on 21 March 1968 (Kosmos 208), and the missions lasted a decade, terminating with Kosmos 1004 (launched from Plesetsk on 5 May 1978).

The use of eight-day missions by the first and second generations of unmanned Soviet PHOTINT spacecraft can be explained through examination of the ground traces of these systems' orbits. Given the Earth's rotation and the orbital parameters set by Soviet mission planners, a general westward drift occurs in the satellite's ground traces, such that the area covered by the vehicle on the eighth day of its mission is the same as that observed during the first day in orbit. A Soviet photo-reconnaissance satellite would thus cover in its entirety the area that lay between those northern and southern latitudes having the same value as the spacecraft's inclination. The politico-military significance which the Soviets assign to recoverable PHOTINT missions that provide regular coverage of all major threat areas (eg the United States, European NATO nations, The Peoples Republic of China, etc) is demonstrated by the continued use of the first generation systems in an area-search mode after the introduction of second generation systems capable of higher resolution coverage. Extended-duration second generation satellites then took up the area coverage mission until 1978, when they were replaced by third generation medium-resolution systems that were boosted to high perigees and employed interlaced coverage.

Third Generation Photographic Reconnaissance Satellites

Seven months after the first extended-duration mission was flown by a second generation spacecraft, the third generation of photo-reconnaissance satellites was launched from Tyuratam (Kosmos 251). Based on the Soyuz manned spacecraft design, this satellite could maneuver in orbit, allowing flexible close-look, high-resolution coverage for the first time. Paired with the second generation extended-duration system, the initial third generation systems gave the USSR the

Annual Days of Soviet Photo-Reconnaissance Satellite Coverage

	1962	1963	1964	1965	1966	1967	1968	1969	1970	1971	1972	1973
Days	22	54	77	122	160	168	198	212	249	244	255	285
Annual %	6	15	21	33	44	46	54	58	68	67	70	78

ability to follow up rapidly on the results of broad area collection with precise point-target imagery.

Third generation PHOTINT systems are thought to resemble the basic Soyuz configuration, with the exception that they use chemical instead of solar batteries. As with the manned Soyuz, the vehicle is probably composed of three large modules: (1) an instrument/propulsion package weighing around 2200 kilograms; (2) a reentry capsule 'shaped like a beehive' that returns the film at the termination of a mission and weighs approximately 2500 kilograms; and (3) a forward module that contains the camera system and perhaps other intelligence gathering apparatus, weighing perhaps 1200 kilograms. The mission life for the initial third generation system (termed Morse Code by Western observers because of its telemetry format) averaged 13 days. The last mission of this type was launched from Tyuratam on 12 February 1974 (Kosmos 632).

Beginning in late 1970, the Soviets introduced a family of third generation PHOTINT systems that remain in service today. Designated Two-Tone satellites, because they do not appear to transmit any telemetry and their recovery beacons consist of a two-tone sequence transmitted on 19.989 MHz, these satellites still perform the bulk of Soviet photo-reconnaissance missions. The three Two-Tone variants provide a spectrum of low-, medium-, and high-resolution coverage of target areas for Soviet analysts and decision-makers.

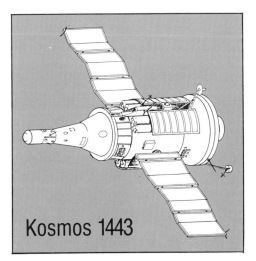

Kosmos 1443

Two-Tone Low-Res system: This third generation variant was inaugurated with a launch from Plesetsk on 27 December 1971 (Kosmos 470). These spacecraft are the most infrequently launched of their generation, averaging two or three missions per year. The satellites are usually placed in a 250×215-kilometer-orbit, inclined at 82.3 degrees, with an 89 minute orbital period. Low-Res is launched exclusively from Plesetsk. The vehicle does not maneuver during its mission life, and its infrequent use, along with the role of broad area search being accomplished by the medium-resolution variant, indicate that this type of spacecraft may have a global geodetic and mapping mission. This supposition is reinforced by the correlation of launches with occurrences of the equinoxes, in what may be an attempt to secure uniform lighting conditions in both the northern and southern hemispheres. Precise mapping is an important military mission in that it allows precise targeting of strategic ballistic weapons with intercontinental ranges. The Low-Res system may be designed to assist efforts to tie together diverse global map grids through simultaneous collections involving photography of terrestrial features and laser measurements of star fields.

Two-Tone Med-Res system: As mentioned above, this space craft replaced the second generation extended-duration variant for broad area missions in 1978. Its first flight was Kosmos 867, launched from Plesetsk on 24 November 1976. Following a 24-hour period in very low-Earth orbit, the satellite is boosted to a second orbit with a perigee above 300 kilometers and an apogee of some 425 kilometers. Med-Res is launched from both Tyuratam (using an inclination of 70.4 degrees) and Plesetsk (using inclinations of 67.1, 72.9and 82.3 degrees). The mission life for this second generation variant is approximately two weeks. Operating from its higher orbit, Med-Res develops ground traces that fall almost mid-way between those of the day before. Given a typical Plesetsk mission inclined at 72.9 degrees, with an orbital period of 89.5 minutes, normal westward drift would cause the ground track of the fourteenth day to coincide with

that of the first day following maneuver to the Med-Res operational orbit. Such interlaced coverage allows the Soviets to conduct extensive search-and-find missions at medium resolution, whereas in the past they had to use low-resolution systems.

Two-Tone Hi-Res system: Kosmos 364—launched from Plesetsk on 22 September 1970—was the first third generation high-resolution photo-reconnaissance satellite. These systems have been launched from both Plesetsk (using inclinations of 67.1, 72.9 and 82.3 degrees) and Tyuratam (using inclinations of 65 and 70.4 degrees). The mission lifetime for these spacecraft is around two weeks. Because Hi-Res systems are often maneuvered to collect against targets of opportunity, they cannot be said to fly a typical orbit. Hi-Res systems have been maneuvered in the past to collect data on various crises and wars, including the 1971 Indo-Pakistani War, the 1973 Middle East War, the 1982 Israeli invasion of Lebanon, the ongoing Iran-Iraq war, the 1983 American invasion of Grenada, the Libyan-backed attempt to overthrow the government of Chad in 1983 and the anti-government insurgency in Somalia in 1984. In addition, these systems provide photography of Afghanistan and northern Pakistan in order to support Soviet efforts to crush the Afghan insurgency against their occupying forces. Finally, Hi-Res variants have been maneuvered to collect information on US and NATO military exercises and on American space program operations, such as launches and landings of the US space shuttle.

The longer missions of the third generation in general, and Hi-Res in particular, are a result of increases in the ability of Soviet photographic equipment to resolve objects on film. Employing longer focal-length optical devices to achieve higher resolution, however, results in a narrower ground swath open to collection at any given time. Thus, in order to cover the same amount of territory, ground traces must be narrowed and time in orbit increased. The Soviets have always operated their PHOTINT systems at the lowest possible perigee, given the restrictions imposed by atmospheric drag, so closer ground tracks and a shorter orbital period are

Annual Days of Soviet Photo-Reconnaissance Satellite Coverage (Continued)

	1974	1975	1976	1977	1978	1979	1980	1981	1982	1983
Days	258	296	306	279	302	303	344	354	341	346
Annual %	71	81	84	76	83	83	94	97	93	95

Source: Jeffrey T. Richelson, *Sword and Shield: Soviet Intelligence and Security Apparatus* (Cambridge, Mass: Ballinger, 1986).

Above: The USSR labeled its prototype space-planes Kosmos 1374, 1445 and 1517. *See also* the photos on pages 55 and 84.

obtained via lowering the apogee. Maneuvering capability is also critical since it allows a stabilized orbit over a given target, giving the satellite the same ground track over a target on successive days.

Estimates of the resolution attainable by the Hi-Res system that have appeared in the open Western press and trade publications tend to be based on analogous comparisons rather than on technical parameters. Thus it is impossible to say with any precision what the third generation camera systems reveal to Soviet analysts. Yet it is clear that the Soviets continue to expend a great deal of effort upon recoverable film systems. Recoverable photo-reconnaissance satellites still comprise the major segment of their PHOTINT program, and Hi-Res missions represent the majority of recoverable launches. It appears, therefore, that the Soviets feel the resolution attainable from their third generation systems to be adequate.

Fourth Generation Photographic Reconnaissance Satellites

The next generation of Soviet PHOTINT systems debuted in 1975, but it probably took five years for the vehicle to reach operational status. The first launch (Kosmos 758, from Plesetsk, on 5 September) exploded in orbit, as did the third attempt. By 1980, however, the Soviets were conducting regular launches of the system from both Tyuratam (using an inclination of 64.9 degrees) and Plesetsk (using an inclination of 67.2 degrees).

The mission lifetime of the fourth generation craft ranges between 23 and 59 days. These extended operations are probably facilitated by the employment of the solar-powered batteries that have been used on the Soyuz vehicle since it was man-rated. Unlike the third generation systems, which deorbit the entire film load in a reentry module, the fourth generation satellite returns its exposed film in four to six recoverable capsules about the size of the Luna 16 soil sample return capsule (see the Luna heading for details).

The film capsules are clustered around the middle of the vehicle where the beehive reentry module was located on the third generation craft. Total weight for a single capsule is approximately 75 kilograms. Behind the capsule cluster is an instrument package that contains the propulsion system and the externally attached solar panels positioned to allow the maintenance of the spacecraft in constant nose-down position with respect to the Earth's surface. The instrument module weighs approximately 2500 kilograms. The remainder of the structure, including the camera package and perhaps containing other means of technical collection, weighs 3650 kilograms. The entire vehicle, assuming four film return capsules and 550 kilograms of fuel, weighs some 7000 kilograms.

The fourth generation craft are high-resolution vehicles that are initially placed in a 175×350-kilometer orbit. Such a low perigee requires weekly maneuvers to maintain orbit due to atmospheric drag. A film capsule may be deorbited in conjunction with the weekly orbital boosts. The capsules use recovery beacons similar to previous generations, but there is some evidence indicating that the capsules are snatched from their descent by aircraft. On other occasions, however, the recovery signals (Morse code TKs transmitted on 19.995 MHz) have persisted much longer than was usually the case with previous generations, indicating Soviet difficulties in locating the smaller capsules. The absence of a recovery beacon associated with the deorbiting of the main vehicle indicates that the Soviets allow these components to burn up upon reentry.

Fifth Generation Photographic Reconnaissance Satellites

In late 1982, the USSR took a step toward acquiring a major qualitative improvement in their photo-reconnaissance program. The fifth generation of PHOTINT systems appears to have progressed beyond the need to return exposed film for subsequent development and analysis on the ground. Until the fourth generation of photo-reconnaissance satellites, the Soviets either had to wait until the end of a mission (upwards of two weeks) in order to discover what had been photographed or make the decision to initiate a premature mission termination, thus losing collection opportunities with respect to whatever rapidly unfolding events might have precipitated deorbiting the vehicle in the first place. The film capsule return system on the fourth generation systems reduced the amount of time between film deliveries, but the Soviets still faced the basic tradeoff between date of information and curtailment of collection capability. The fifth generation appears to be what is known as a real-time system. It sends imagery back to ground stations in digital form by means of a relay conducted via a communications satellite in geosynchronous orbit. This new capability should augment the USSR's ability to monitor crises and conflicts around the world, target its long-range weapon systems and verify arms control agreements. In addition, the elimination of a film return system has resulted in a major improvement in mission lifetimes. The three missions that have been identified with the fifth generation by Western experts are described below.

Kosmos 1426

Launch Vehicle: A-2
Launch Site: Tyuratam
Launch Date: 28 December 1982
Recovery Date: 5 March 1983
Total Weight: 7000 kilograms
Apogee: 356 kilometers
Perigee: 202 kilometers
Inclination: 50.53
Period: 90.09 minutes

Initially, Western observers thought that this spacecraft had some relationship to the Soviet manned space program. Since the plane of its initial orbit (337×198 kilometers) was almost coincident with that of the Salyut 7 space station, some analysts predicted that a rendezvous of some type was in the offing. On 3 January 1983, however, the vehicle was maneuvered into a 351×205-kilometer orbit, which is more consistent with a photo-reconnaissance mission. During its 67-day mission, Kosmos 1426 had ten major changes of orbit. Throughout these maneuvers the spacecraft maintained orbital parameters consistent with a photo-reconnaissance mission (200×300–400 kilometers), and the point of perigee was maintained so that the satellite operated in daylight conditions. No other missions of this type were conducted for the remainder of 1983, indicating the developmental nature of the program.

Kosmos 1552

Launch Vehicle: A-2
Launch Site: Tyuratam
Launch Date: 14 May 1984
Recovery Date: 3 November 1984
Total Weight: 7000 kilograms
Apogee: 322 kilometers
Perigee: 182 kilometers
Inclination: 64.93 degrees
Period: 89.54 minutes

The next mission in the fifth generation development program was not launched until late spring 1984. Like its predecessor, Kosmos 1552 maintained a perigee that was con-

siderably higher than those associated with the fourth generation high-resolution missions. The vehicle also adjusted its orbit in a manner similar to Kosmos 1426 in order to prevent orbital decay, conducting 12 major maneuvers over the course of its orbital life. Unlike the earlier mission, however, Kosmos 1552 deliberately lowered its orbit in order to stabilize its ground traces and facilitate collection of data probably associated with the Iran-Iraq War. The biggest difference between the two flights, however, was the lifetime of the Kosmos 1552 mission, which lasted 173 days. While Kosmos 1426 demonstrated a lifetime 12 percent greater than the longest fourth generation flight, its successor revealed a 250 percent leap in capability.

Kosmos 1643

Launch Vehicle: A-2
Launch Site: Tyuratam
Launch Date: 25 March 1985
Recovery Date: 18 October 1985
Total Weight: 7000 kilograms
Apogee: 294 kilometers
Perigee: 223 kilometers
Inclination: 64.77 degrees
Period: 89.68 minutes

This mission had many of the features of the earlier fifth generation launches, although it appears that Tyuratam was the launch site of preference for this generation

Above: **A display model of Kosmos 782 is here shown against a backdrop of explanatory charts. Note the open payload module. This vehicle carried US fauna.**

of PHOTINT systems—a decision that reversed a decades-long trend. Kosmos 1643 maintained the now-familiar high perigee, but it reverted to the orbital behavior exhibited by Kosmos 1426, in that it did not stabilize its ground track so as to conduct close-look collections of data on point targets. Like its predecessors in 1982–83 and 1984, the fifth generation vehicle set a new lifetime record, operating for 207 days before recovery in late 1985. The lack of any attempt deliberately to lower the vehicle's perigee for point collection has led some Western experts to speculate that the fifth generation's resolution is such that, unlike its predecessor, Kosmos 1552, it is only suited for global area coverage.

Photographic Reconnaissance Missions Conducted By Manned Orbital Platforms

There is substantial evidence to indicate that the Salyut space stations have had a major reconnaissance mission on at least several occasions in the past. Details concerning the Salyut program can be found in Appendix I. This subsection is concerned with these vehicles only to the extent that they have been employed for reconnaissance missions.

Early in the Salyut program it appeared that the Soviets were developing an operational profile that featured alternating civilian and military missions. Salyut 3 and 5 were both placed in lower orbits than Salyut 1 and 4, which could have facilitated PHOTINT collection. Unlike the other Salyut crews, which were mixed civilian and military, Salyut 3 and 5 had all-military crews, and when they entered the space stations at the start of their respective missions the telemetry transmissions were switched to military channels. The telemetry was formatted similarly to that employed by first generation photo-reconnaissance satellites. Following the departure of both crews from their stations, capsules were ejected, deorbited and recovered. Some Western experts have speculated that these recovery capsules contained films from a Cassegrainian instrument used for PHOTINT collection. This camera system reportedly has a focal length of 33 feet, and is capable of providing 12- to 18-inch resolution from orbits such as those in which Salyut 3 and 5 operated.

The second generation of Salyut space stations did not adhere to the first generation's strict civilian/military dichotomy. The example of Salyut 6, however, demonstrates the overlap that appears to occur between civilian and military missions in so many aspects of the Soviet space program. During the lengthy Salyut 6 mission, some 13,000 photographs of the Earth's surface were taken by crew members using KATE-140 and MKF-6M cameras. Despite its distinctly unmartial name, the Mir station could also prove to be a useful platform for photo-reconnaissance and other types of intelligence collection.

Electronic Intelligence Satellites

The purpose of electronic intelligence satellites is to locate emitting radars and to collect data concerning their operational parameters (*eg* pulse repetition frequencies, signal strength, operating frequencies, antenna types, side lobe characteristics, etc). In many cases, positional and operational data concerning radars located well inland in the continental United States and Western Europe is only accessable via satellite. In addition, Soviet ELINT satellites can be utilized during US, NATO and Chinese exercises in order to collect data concerning mobile radars and other emitters.

Space-based ELINT collectors provide critical information in the game of electronic defense and countermeasures in which the opposing sides strive to deny potential adversaries the ability to penetrate their air defenses or blind their early warning mechanisms, while attempting to give precisely such capabilities to their own offensive forces. Soviet ELINT satellites help collect data that facilitate the jamming and/or physical destruction of radars that could be used in time of war by an adversary to provide attack warning and assessment; to protect vital targets such as airfields, command centers and special weapons storage sites; and to track or control satellites. By supplying information about mobile radars and other tactical emitters, Soviet ELINT satellites contribute to construction of the electronic order of battle, a critical intelligence data base used in targeting enemy forces during conflict. Finally, by providing the Soviet political and military leadership with information concerning the disposition of an adversary's radars during a crisis, Soviet ELINT spacecraft can provide timely warning of hostile intent.

This section describes the evolution of Soviet ELINT satellites that provide worldwide coverage (although they appear to be directed primarily against the North American continent). The Soviets have also deployed an ELINT satellite dedicated to ocean surveillance missions. The electronic ocean surveillance satellite, or EORSAT, is covered in the next section on Soviet ocean reconnaissance satellites.

Ellipsoid Elint Satellites

The first Soviet ELINT satellite was Kosmos 148, launched on 16 March 1967 from Plesetsk on a B-1 booster. This series of satellites was dubbed ellipsoid because of the vehicles' shape. The limited lifting power of the B-1 booster probably explains the relatively small total weight of these craft (about 400 kilograms). The body of the spacecraft was about 1.8 meters long and roughly 1.2 meters wide.

With the exception of some possible developmental flights, all of the 64 launches in the series were conducted from Plesetsk. A typical orbit was 170×300-500 kilometers, with an inclination of 71 degrees and a period of around 92 minutes. The ellipsoid ELINT satellites were solar powered and apparently operated in a store-and-dump mode (that is, they collected and stored data concerning emitting radars on each revolution and down-linked this information to ground stations in the Soviet Union when they could establish line-of-sight communications with them). The last mission in this series was Kosmos 919, which was launched on 18 June 1977.

Cylindrical Elint Satellites

Kosmos 189, launched on 30 October 1967, inaugurated the second series of Soviet ELINT ferrets categorized by the vehicle shape. This series was also launched exclusively from Plesetsk, but it was inserted into orbit by the more powerful C-1 booster which allowed greater payload weight. Cylindrical ELINT spacecraft weighed around one metric ton and were two meters long and one meter wide. Paddle-shaped solar panels, extended at an angle to the vehicle axis, supplied power for the instrumentation package.

Cylindrical ELINT ferrets were deployed in constellations of four vehicles, separated from each other by 45 degree planes. A typical orbit for this type of satellite was 550×525 kilometers, with a 74 degree inclination and an orbital period of 95 minutes. The program comprised a total of 38 launches, the last of which occurred on 31 March 1982 (Kosmos 1345).

First Generation Heavy Elint Satellites

With the launch of Kosmos 389 on an A-1 booster from Plesetsk in late 1970, the Soviets began assembling a constellation of satellites that would provide the backbone of space-based ELINT collection for the next 13 years. These craft were referred to as 'heavy' by Western specialists because they significantly exceeded their predecessors in both size (five meters long and 1.5 meters wide) and weight (3800 kilograms). The constellation consisted of six satellites in circular orbits of 620 to 660 kilometers, inclined at 81.2 degrees to the equator, with orbital periods of approximately 95 minutes, and occupying orbital planes spaced at 60 degrees. Some 35 satellites were launched as part of this program between 1970 and 1983. The last launch occurred on 16 February 1983 (Kosmos 1441).

Second Generation Heavy Elint Satellites

In 1978 the Soviets began inserting satellites into orbits of roughly the same altitude as the first generation of heavy ELINT ferrets, but which were inclined at 82.5 degrees (as opposed to the 81.2 degree inclination employed theretofore by the heavy ELINTs). Furthermore, the new spacecraft were launched on the F-2 vehicle, as opposed to the A-1.

In 1981 the Soviets only replaced three of their first generation heavy ELINT ferrets, the lowest number of replacement launches in 12 years. Given a nominal 600-day orbital life for the first generation heavies, a minimum of four annual replacements is required in order to maintain global coverage. The year 1982 saw the resumption of four replacement launches (three within 75 days), but when the maneuvers had been completed it was not clear whether first generation craft had been used for replacement of the aging ferrets. By the end of 1982 the traditionally tight 60-degree spacing for the orbital planes of the satellites in the heavy ELINT constellation had clearly been disrupted.

Additional launches into the 82.5-degree inclination orbits during 1983 brought the total of what appeared to be a second generation of heavy ELINT satellites to eight. By 1984 it was clear that the first generation had been phased out in favor of a successor constellation placed in circular orbits at altitudes of 635–665 kilometers. The new inclination is probably the result of switching to the newer, more accurate F-2 booster. Six second generation ELINT payloads were orbited in 1985, bringing the total by year end to 17. The constellation will probably ultimately achieve 59.5-degree separation between orbital planes. The second generation of heavy ELINT satellites have a location accuracy for active emitters in the ten meter range, and deliver collected data to their ground stations in a record-playback mode. The Soviets also appear to be placing two satellites into each orbital plane, perhaps as orbital spares or in order to facilitate geolocation of emitter signals.

Kosmos 1603-Type Elint Satellites

On 28 September 1984 the Soviets launched Kosmos 1603 from a pad at Tyuratam on a D-1e booster. After three major maneuvers that included two inclination changes, the satellite reached a final orbit of 856×850 kilometers, with an inclination of 71 degrees. This orbit, while lower than that of the second generation of heavy ELINT systems, is still high enough to allow pole-to-pole coverage. Furthermore, the orbit repeats every 14 revolutions—thus covering every target in its path every 24 hours. The orbital parameters ensure that the system passes over the continental United States frequently.

In early 1985 the trade journals began labeling Kosmos 1603 a new type of Soviet ELINT satellite. Its weight was given as 4.5–6 metric tons, and its size as between 33 and 35 square meters. Three more missions of this type were launched in 1985, but the results of the program to date must be disappointing to Soviet mission planners. Kosmos 1656, launched on 30 May 1985 from Tyuratam, was placed in an orbital plane separated from its predecessor by 45 degrees to the east, but it attained a slightly lower orbit (861×801) with a resulting difference in inclination of .1 degree.

On 22 October Kosmos 1697 achieved an almost identical orbit to that of Kosmos 1603, but it employed a completely different launch profile (and perhaps a different booster type) to do so. In addition, Kosmos 1697's orbital plane is between that of its two predecessors (only 11 degrees from Kosmos 1656), apparently disrupting what many Western observers had expected to be a four-vehicle constellation, with orbital planes separated by 45 degrees. In December four objects were detected with the same perigee as Kosmos 1697, but at apogees up to 300 kilometers higher.

Finally, Kosmos 1714 was launched on 28 December. An apparent failure of its propulsion system left the spacecraft stranded in a 853×163-kilometer orbit, inclined at 71 degrees. This satellite also released four associated objects, and they were boosted to apogee some 250 kilometers above the main vehicle. Had Kosmos 1714 reached its proper orbit, it would have confirmed the 45-degree spacing that Soviet mission planners had probably originally wanted to achieve by the end of 1985.

Ocean Reconnaissance Satellites

For the past 29 years the Soviets have conducted a spaceborne monitoring program designed to detect, locate and classify potentially hostile naval surface vessels. Such information would be extremely useful during a conflict, allowing precise targeting of hostile naval surface groups for anti-ship missiles launched from Soviet air, surface or subsurface platforms. The magnitude of the threat caused by these satellites to US carrier battle groups was demonstrated when the US Department of Defense cited Soviet ocean reconnaissance spacecraft as one of the primary justifications for the development of an American anti-satellite capability.

Two types of satellites compose the space-based component of the Soviet ocean surveillance program. The radar ocean reconnaissance satellite system (dubbed RORSAT by US defense analysts) is designed to locate large surface vessels and concentrations of smaller ships via active illumination by radar energy. The ELINT ocean reconnaissance satellite system (or EORSAT) locates hostile fleet elements by collecting the emissions of their radar and communication systems. This section describes the physical and operational characteristics of these two spacecraft and provides a tabular launch history of their missions.

The first vehicle test for RORSAT was conducted in 1967. The spacecraft is always launched from Tyuratam on an F-1m booster. RORSAT's operational parameters generally include almost circular orbits with perigees between 250 and 260 kilometers, inclined at 65 degrees, with orbital periods of 89.5 minutes. The spacecraft is composed of the vehicle itself and a non-separating third stage of the booster. An ion engine is attached to the spent booster stage for orbital maintenance (the choice of a mean 255-kilometer altitude represents a trade-off between the power of RORSAT's sensor and the probability of detecting objects on the ocean's surface). Forward of the F-1m launch vehicle's third stage is an instrument module that carries the RORSAT's electronic and attitude control instrumentation. Communications equipment in the instrument module broadcasts at 16 MHz. Beyond the instrument module is the radar antenna, which has been described in open Western sources as either a planner array or a slot-type antenna. The antenna is attached to the main body of the vehicle, which contains a boost stage for placing the power supply module in a higher orbit at the end of mission life. Two antennas protrude from the forward section of the main body, which contains a boost-stage electronics module that transmits at 19.542 Mhz. The power supply is mounted on the nose of the spacecraft. It is a Topaz thermionic reactor-convertor which supplies 10 KW of power.

RORSAT has been one of the most error-plagued elements of the Soviet space program. The majority of difficulties stems from its reliance upon nuclear power and the resulting danger of contamination, should a RORSAT deorbit in such a way as to allow the reactor core to survive reentry. From the beginning of the program, Soviet mission planners have sought to prevent just such an occurrence by boosting the reactor to a 900- to 1000-kilometer orbit, from which it will not decay for hundreds of years. (The other components of the vehicle remain in low-Earth orbit, from which they decay naturally in a few days or weeks.) The 16th RORSAT (Kosmos 954), however, failed to execute the lofting maneuver for its reactor at the end of its mission life. In early November 1977 the spacecraft entered a decaying orbit, and, following the loss of attitude control in early 1978, deorbited over northern Canada. Radioactive debris was scattered over a large area of the essentially unpopulated Canadian tundra. A joint US-Canadian clean-up effort decontaminated most of the effected area, and the Soviet RORSAT program entered a two-year hiatus.

Kosmos 1176, the next RORSAT launch, demonstrated the solution chosen by the Soviets to meet the problem of premature reentry of the reactor module. Reasoning that the reactor housing had shielded the radioactive fuel core from the effects of reentry, Soviet specialists redesigned the vehicle's boost stage to permit the ejection of the core so as to maximize the chances that radioactive debris would not reach the surface of the planet. The next six RORSATs following Kosmos 1176 performed the tri-partite separation and the reactor lofting maneuver, followed by the ejection of the core once the reactor had reached an altitude of roughly 900 kilometers.

The Soviets had the chance to test their modified nuclear safety procedures when, in late 1982, Kosmos 1402 encountered end-of-mission difficulties. Vehicle separation yielded only two segments rather than the intended three, and when it apparently became impossible to boost the reactor to a higher orbit, Soviet mission controllers ejected the core. The reactor deorbited over the Indian Ocean on 24 January 1983. Two weeks later, the reactor core reentered the atmosphere over the south Atlantic.

Very little of the radioactive debris from Kosmos 1402 apparently reached the planet's surface, but the Soviets could not have been pleased by the second mission failure in five years. Another hiatus in ROR-SAT launches followed, this time lasting a year. Evidently the Soviets decided to add further modifications to the vehicle, the mission control apparatus or both. These changes have not been apparent since the resumption of RORSAT flights in 1984, and may not be until another lofting failure occurs.

In a larger sense, even successful boostings of reactors into orbit may be dubious achievements. Western specialists have pointed out that the debris from Kosmos 942 contained radioactive isotopes of Uranium-235 (93 percent enriched). While the parking orbits for the reactor housings and cores assure reentry in 300 to 1000 years, the half-life of U-235 is over 70,000 years. When the lofted RORSAT elements eventually do reenter, they will still pose a major threat of contamination, especially those elements in the 15 reactors boosted before Kosmos 954, which have not ejected their cores.

EORSAT launches began in 1974. The F-1s booster is employed to place the vehicle in an eccentric orbit measuring 115×405 kilometers. Within 45 minutes of achieving this initial orbit, the vehicle employs a booster engine to reach its operational circular orbit of 435 kilometers, with a 65-degree inclination and a 93.3-minute period. This orbit is maintained against drag via the use of an ion-thruster engine. EORSATs usually operate in pairs, spaced 180 degrees apart, so that both vehicles create identical (although not simultaneous) ground traces. EORSAT is a solar-powered system and thus has not been subject to failures with consequences of the proportions that have afflicted its sister system. End-of-life EOR-SATs are either allowed to decay naturally or

are boosted into higher orbit and then fragmented.

RORSATs and EORSATs also appear to operate together when an ocean reconnaissance constellation contains both types of satellites (EORSATs tend to have much longer operational lives than RORSATs, and, as we have seen, RORSAT has suffered periodic lengthy absences from duty). RORSATs usually have an initial planar separation from EORSATs of 145 degrees, which tends to expand to about 175 degrees by the end of a mission, due to orbital perturbations. Resolution of RORSAT has been reported in the Western press to be destroyer-sized objects in calm seas and carrier-sized objects (or clusters of smaller vessels) in rough seas. EORSAT frequencies and geolocation accuracies have not been reported in the open literature. Targeting data derived from Soviet ocean reconnaissance satellites can be down-linked in real-time to platforms capable of receiving such transmissions. They can also be passed in a less timely manner through ground reception sites in the USSR to deployed forces without means of direct reception. Table 3 provides the history of RORSAT and EORSAT launches.

Early Warning Satellites

Perhaps the most problem-plagued and expensive Soviet military space program has been the effort to establish an operational network of early warning satellites. Since the late 1960s, the USSR has sought to effect a space-based capability to provide roughly 30 minutes of tactical warning of US ICBM and SLBM launches, as well as a determination of the areas from which the attack was launched. Early warning satellites detect the exhaust plumes of ballistic missiles as they top the clouds above their launch points by scanning these geographic areas in the infrared spectrum. Such 'tactical' warning is vital because it would provide time for emergency dispersal of key leadership elements, the launching of endangered strategic missiles prior to the impact of in-coming warheads ('launch-on-warning'), and activation of strategic defenses.

The Soviets began testing satellites for an early warning role in 1967. Probably in recognition of their traditional weakness in placing systems in geosynchronous orbit, Soviet mission specialists turned to the highly elliptical semisynchronous orbit previously developed for their Molniya series of communications satellites. These spacecraft have a highly eccentric orbit of 400×40,000 kilometers. This gives them a roughly nine-

hour period close to apogee when they can communicate easily with Earth. (See the entry under the Molniya heading for greater details on the basic orbit and applications of this type of satellite.)

The first Molniya test was Kosmos 174, launched from Tyuratam on 31 August 1967 on an A-2e booster. Following a second flight from Tyuratam the next year, the Soviets switched to Plesetsk for the rest of their Molniya early warning satellite launches. It was the series of Plesetsk launches, beginning in 1972, which alerted Western observers to the differences between the early warning orbits and those associated with communications applications. Specifically, the arguments of perigee for the early warning craft were discovered to be some 315 degrees, as opposed to the 280 degrees usually associated with the Molniya communications network. Soviet early warning satellites are usually inclined at 62.8 degrees, with a 12-hour orbital period, and have a total system weight of 1800 kilograms.

The Molniya orbit can satisfy the basic early warning requirements for the USSR: Satellites must attain an altitude from which the vehicle can simultaneously view ICBM fields in the western and mid-western United States—or the ballistic missile submarine (SSBN) patrol areas in the eastern Atlantic, western Pacific and northern Indian Oceans—and communicate in real-time with key Soviet ground stations. Such orbits are accomplished by placing the Molniya ascending node some 25 degrees east of its usual position for communications applications (115 degrees west and 65 degrees east).

Between 1972 and 1980 the Soviet early warning satellite program was essentially experimental. A constellation of three satellites was maintained during this period, with each vehicle separated from the others by 80 degree orbital planes. Such a configuration could not have maintained continuous coverage of US ICBM fields and SSBN patrol areas. In 1980, a fourth satellite was added to the constellation, and in 1981 it became clear that the USSR intended to establish an operational constellation of some nine satellites spaced at 40-degree planar differences and following each other onto station at continuous two-hour-and-40-minute intervals.

In 1982 Soviet mission planners put the whole early warning satellite constellation into a series of maneuvers that resulted in a major shift in the constituent satellites' ascending nodes. The new orbits had ascending nodes some 30 degrees east of their previous positions, ending up at approximately 120 degrees west and 300 degrees east. This change resulted in better vantage points for scanning the continental United States, the

Atlantic Ocean and the People's Republic of China.

But the Soviets experienced major difficulties in establishing and maintaining their early warning satellite network in the years following 1982. In 1983, out of three attempted launches two were failures, bringing the overall failure rate for the program to 50 percent. The year 1984 saw a record seven launches in an all-out Soviet effort to establish a fully operational early warning constellation. By year's end, however, at least one satellite was not functioning. The next year was a virtual repetition of 1984, with seven early warning launches in five months. Once again, however, the strenuous Soviet efforts could not assure a completely functional constellation—at least two satellites were not functioning properly by the close of 1985.

Soviet fortunes concerning the often ill-fated early warning satellite program reveal a mixture of strength and weakness. The ability to replace numerous malfunctioning satellites in a short period of time, as revealed in 1984 and 1985, indicates that the USSR could rapidly overcome catastrophic mechanical failure of their early warning spaceborne network or replace them in the event of loss due to hostile action. On the other hand, the need for frequent replacement indicates an unacceptable failure rate and some serious technical weaknesses. Furthermore, the apparent failure even to attempt to establish a geosynchronous warning network (with the possible exception of Kosmos 775, which might have tested a geosynchronous observation capability in the mid-1970s) would seem to relegate the Soviets to the cumbersome Molniya orbit for at least the near future.

Military Communications Satellites

The Soviets maintain three tiers of communications satellites, all of which have some military function: low-Earth orbit, semisynchronous and geosynchronous. Those systems that are dedicated to military use or have partial military functions are described in this section. More details for many of the systems discussed below can be found under separate system headings (eg, Raduga).

Covert store-dump communications satellites: Beginning in 1967, the Soviets deployed what appear to be communications satellites that are used for clandestine intelligence-related missions. Three of these vehicles are usually operational at any given

time. They are inserted into orbital planes, spaced 120 degrees apart, in basically circular orbits at 800 kilometers altitude, inclined at 74 degrees, with orbital periods of around 100 minutes. Launched from Plesetsk on C-1 boosters, these spacecraft weigh approximately 750 kilograms and provide near-real-time communications for Soviet intelligence operatives worldwide.

Communications satellites supporting theater-level and other deployed forces: Other Soviet store-dump communications systems operate in circular orbits of around 1500 kilometers in constellations of up to thirty satellites. These satellites are thought to provide real-time tactical communications for individual theaters of military operations, and near-real-time links between widely dispersed Soviet military forces (eg, naval flotillas) and central Soviet command entities. These systems may also play a role in communications between Soviet land-based or sea-based mobile missile units.

The C-1 booster is employed from Plesetsk to place a package of eight of these small (around 40 kilograms) spacecraft into orbits inclined at 56 degrees, with periods of about 115 minutes. The octuple payload is carried to the highest operational altitude achievable with the C-1 in order to create a belt of satellites that facilitates communications via a reduction of the delays that occur when the vehicles are out of line-of-sight from various terrestrial users. The C-1 octuplet launches are the largest Soviet multi-payload efforts. Generally, two to three launches occur per year.

Molniya military communications: Both Molniya 1 and Molniya 3 communications satellites appear to have some military functions (see these entries under the Molniya heading for the spacecrafts' operational parameters). The Molniya 1 system is generally believed by Western experts to be dedicated to Soviet governmental and military communications. Four of the Molniya 1 vehicles appear to transmit only during the Asian apogee climb, and these transmissions differ from other Molniya-related emissions to such an extent that some Western specialists have labeled them military in nature.

The Molniya 3 system appears to have Soviet responsibility for the Hot Line link between Washington and Moscow for crisis communications. During its climb to apogee over North America, a Molniya 3 seems to maintain a channel for transmission between Hot Line terminals at Vladimir in the USSR and Ft Detrick, Maryland. In 1984, the US and the USSR agreed to upgrade the Hot Line by increasing the word-per-minute rate and establishing the means for data-linking

Chameleons: *Above* is a Kosmos satellite based on a modified Soyuz spacecraft which itself was based on the diving bell—the Soviet 'incremental approach.'

pictures and other graphics between the two capitals. The Molniya 3 system will probably be upgraded to accommodate these improvements.

Geosynchronous military communications satellites: There is no clear evidence of Soviet military geosynchronous communications in the open literature. The USSR announced in the mid-1970s that four Gals military communications satellites would be placed in geostationary orbits in 1979, but by the end of 1985 no such launches had taken place. The Soviets may, however, be placing Gals transponders on their Raduga geosynchronous communications systems, since these spacecraft have recently been placed in the slots originally earmarked for Gals (see the Raduga entry for operational and orbital parameter data on this satellite).

Military Navigation Satellites

Navigation satellites allow precise determination of geographic locations on Earth. Since the late 1960s the Soviets have operated navigation satellites for both military and civilian applications. The military program appears to have

Rorsat and Eorsat Missions

Kosmos #	Launch Date	Launch Site	Launch Vehicle	Weight (kgs)	Apogee (kms)	Perigee (kms)	Inclination (dgrs)	Period (mins)
198	27/12/67	TT	F-1-m	4540	281	265	65	89
209	22/03/68	TT	F-1-m	4540	282	250	65	89
367	03/10/70	TT	F-1-m	4540	280	250	65	89
402	01/04/71	TT	F-1-m	4540	279	261	65	89
469	25/12/71	TT	F-1-m	4540	276	259	65	89
516	21/08/72	TT	F-1-m	4540	277	256	65	89
626	27/12/73	TT	F-1-m	4540	280	257	65	89
651	15/05/74	TT	F-1-m	4540	276	256	65	89
654	17/05/74	TT	F-1-m	4540	277	261	65	89
699	24/12/74	TT	F-1-s	4150	454	436	65	93
723	02/04/75	TT	F-1-m	4540	277	256	65	89
724	07/04/75	TT	F-1-m	4540	276	258	65	89
777	29/10/75	TT	F-1-s	4150	456	437	65	93
785	12/12/75	TT	F-1-m	4540	278	259	65	89
838	02/07/76	TT	F-1-s	4150	456	438	65	93
860	17/10/76	TT	F-1-m	4540	278	260	65	89
861	17/10/76	TT	F-1-m	4540	280	256	65	89
868	26/11/76	TT	F-1-s	4150	457	438	65	93
937	24/08/77	TT	F-1-s	4150	457	438	65	93
952	16/09/77	TT	F-1-m	4540	278	258	65	89
954	18/10/77	TT	F-1-m	4540	277	259	65	89
1094	18/04/79	TT	F-1-s	4150	457	437	65	93
1096	25/04/79	TT	F-1-s	4150	457	439	65	93
1167	14/03/80	TT	F-1-s	4150	457	433	65	93
1176	29/04/80	TT	F-1-m	4540	265	260	65	89
1220	04/11/80	TT	F-1-s	4150	454	432	65	93
1249	05/03/81	TT	F-1-m	4540	264	251	65	65
1260	21/03/81	TT	F-1-s	4150	447	428	65	93
1266	16/04/81	TT	F-1-m	4540	267	248	65	89
1286	04/08/81	TT	F-1-s	4150	444	431	65	93
1299	24/08/81	TT	F-1-m	4540	265	251	65	89
1306	14/10/81	TT	F-1-s	4150	443	429	65	93
1337	11/02/82	TT	F-1-s	4150	446	428	65	93
1355	29/04/82	TT	F-1-s	4150	446	428	65	93
1365	14/05/82	TT	F-1-m	4540	265	248	65	89
1372	01/06/82	TT	F-1-m	4540	264	250	65	89
1402	30/08/82	TT	F-1-m	4540	264	251	65	89
1405	04/09/82	TT	F-1-s	4150	445	427	65	93
1412	02/10/82	TT	F-1-m	4540	263	250	65	89
1461	07/05/83	TT	F-1-s	4150	447	429	65	93
1507	29/10/83	TT	F-1-s	4150	436	426	65	93
1567	30/05/84	TT	F-1-s	4150	441	431	65	93
1579	29/06/84	TT	F-1-m	4540	264	250	65	89
1588	07/08/84	TT	F-1-s	4150	446	426	65	93
1607	31/10/84	TT	F-1-m	4540	263	251	65	89
1625	23/01/85	TT	F-1-s	4150	392	120	65	89
1646	19/04/85	TT	F-1-s	4150	444	428	65	93
1670	01/08/85	TT	F-1-m	4540	264	252	65	89
1677	23/08/85	TT	F-1-m	4540	263	251	65	89
1682	19/09/85	TT	F-1-s	4150	443	429	65	93

Notes on Rorsat and Eorsat Missions

Kosmos	Mission	Remarks
198	Rorsat	Vehicle test. Reactor lofted to higher orbit on 29/12/67.
209	Rorsat	Vehicle test. Reactor lofted to higher orbit on 28/3/68.
367	Rorsat	Probable failure. Reactor lofted soon after launch on 03/10/70.
402	Rorsat	First operational mission. Reactor lofted to higher orbit on 09/4/71.
469	Rorsat	Reactor maneuvered to higher orbit on 04/1/72.
516	Rorsat	Reactor maneuvered to higher orbit on 21/9/72.
626	Rorsat	Reactor maneuvered to higher orbit on 11/2/74.
651	Rorsat	Initial vehicle of first Rorsat pair. Reactor lofted on 25/7/74.
654	Rorsat	Paired with Kosmos 651. Reactor moved up on 30/7/74.
699	Eorsat	First Eorsat mission. Exploded by ground control on 17/4/75.
723	Rorsat	Initial vehicle of Rorsat pair. Reactor lofted on 16/5/75.
724	Rorsat	Paired with Kosmos 723. Reactor lofted on 12/6/75.
777	Eorsat	Exploded by ground control in January 1976.
785	Rorsat	Probable failure. Reactor lofted after one day of operation on 13/12/75.
838	Eorsat	Initial vehicle of first Eorsat/Rorsat pair. Exploded by ground control circa June/July 1977.
860	Rorsat	Paired with Kosmos 838. Reactor moved up on 10/11/76.
861	Rorsat	Replaced Kosmos 860 and paired with Kosmos 838 and Kosmos 868 to form the first Rorsat/Eorsat triplet. Reactor lofted on 20/12/76.
868	Eorsat	Decayed 8/7/78.
937	Eorsat	Paired with Kosmos 952. Decayed 19/10/78.
952	Rorsat	Reactor maneuvered to a higher orbit on 8/10/77.
954	Rorsat	First separation failure. Decayed over northern Canada on 24/1/78.
1094	Eorsat	Decayed 7/11/79.
1096	Eorsat	Paired with Kosmos 1094. Decayed 24/11/79.
1167	Eorsat	Paired with Kosmos 1176 and 1220. Destroyed by ground control 15/7/81.
1176	Rorsat	First improved version of reactor lofting technique, which was demonstrated on 10/9/80.
1220	Eorsat	Placed in an elliptical orbit and destroyed by ground control on 20/6/82.
1249	Rorsat	Paired with Kosmos 1260 and Kosmos
		1266. Reactor lofted on 19/6/81.
1260	Eorsat	Placed in an elliptical orbit and destroyed by ground control on 8/5/82.
1266	Rorsat	Probable malfunction led to reactor being moved up in less than a week from launch on 29/4/81.
1286	Eorsat	Paired with Kosmos 1299 and 1306. Destroyed by ground control on 29/9/82.
1299	Rorsat	Probable malfunction caused reactor lofting maneuver on 6/9/81.
1306	Eorsat	Destroyed by ground control on 12/7/82.
1337	Eorsat	Probable propulsion system failure led to reentry and destruction on 25/7/82.
1355	Eorsat	Operated with two Rorsat pairs (Kosmos 1365&1372, and 1402&1412). Destroyed by ground control on 8/8/83.
1365	Rorsat	Paired with Kosmos 1372. Reactor moved up on 27/9/82.
1372	Rorsat	Reactor maneuvered to higher orbit on 11/8/82.
1402	Rorsat	Paired with Kosmos 1412 and Kosmos 1355&1405 Eorsat pair. Reactor separated from vehicle but failed to boost to higher orbit, reentering circa January/February 1983.
1405	Eorsat	Destroyed by ground control on 20/12/83.
1412	Rorsat	Reactor maneuvered to higher orbit on 10/11/82.
1461	Eorsat	Placed in elliptical orbit and destroyed by ground control on 11/3/85.
1507	Eorsat	Paired with Kosmos 1461.
1567	Eorsat	Set a record for 538 days of operational life.
1579	Rorsat	Operated with Kosmos 1567&1588 Eorsat pair. Reactor moved up on 28/9/84.
1588	Eorsat	Participated in two Eorsat/Rorsat triplets.
1607	Rorsat	Operated with Kosmos 1567&1588 Eorsat pair. Reactor lofted on 31/1/85.
1625	Eorsat	Probable propulsion system failure stranded vehicle in parking orbit from which it decayed on 25/1/85.
1646	Eorsat	Formed Eorsat triplet with Kosmos 1567 and Kosmos 1682, which operated with Kosmos 1670&1677 Rorsat pair.
1670	Rorsat	Reactor maneuvered to higher orbit on 22/10/85.
1677	Rorsat	Reactor maneuvered to higher orbit on 23/10/85.
1682	Eorsat	Still in operation by end of 1985.

Sources: *Soviet Space Programs: 1976–1980* (U.S. Senate, Committee on Commerce, Science, and Transportation, Washington, D.C.: U.S. Government Printing Office) Parts 1 (December 1982) and 3 (May 1985); and Nicholas Johnson, *The Soviet year in space*, years 1981–1985 (Colorado Springs, CO: Teledyne Brown, 1982–1986).

Kosmos 144

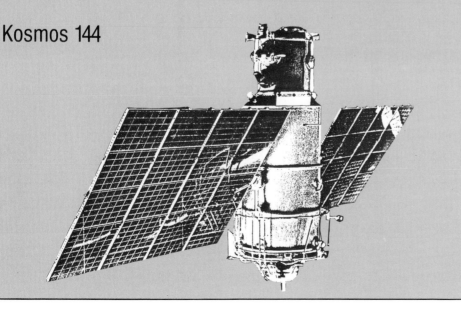

been initiated in order to support Soviet missile submarines. The accuracy of a submarine-launched ballistic missile depends on how precisely the launcher can locate the position of the launch platform vis-a-vis that of the intended target. Other military applications include providing targeting updates for aircraft and missiles in flight, location information updates for naval and ground forces and fire control updates for ground and naval artillery bombardment. The Soviet civilian navigation satellite network, designed primarily to provide locational data to the USSR's merchant marine and fishing fleets, is covered under the Civilian Kosmos entry.

First Generation (Transit-Type) Soviet Navigation Satellites

Soviet navigation satellite networks appear to be essentially clones of American systems. In the early 1960s the United States developed the Transit satellite network in order to support its Polaris SSBNs. On 23 November 1967 the USSR launched its first experimental navigation satellite (Kosmos 192) from Plesetsk on a C-1 booster. Three years later, an operational network was established.

The first generation Soviet navigation satellites operate in constellations of six, with a planar separation of 30 degrees. The spacecraft are placed in circular low-Earth orbits, inclined at 83 degrees, with orbital periods of approximately 105 minutes. Total vehicle weight is around 680 kilograms. Average system life is around 15 months. Operating frequencies are 149.99MHz and 399.97 MHz. These characteristics bear a striking similarity to the US Transit system.

A user determines his location through measurement of the Doppler shift between the two broadcast frequencies, which is calculated by combining ephemeris data broadcast by the spacecraft with satellite tracking data from ground stations. The resulting computation yields an estimated accuracy of 50 meters in two dimensions. Unlike the United States, which now makes most of its navigation satellite data available to any platform capable of receiving it, the USSR does not make its Transit-type network available to general users.

The value of navigation satellites to the Soviet military was indicated in 1982 during a major strategic weapons exercise. In one seven-hour period on 18 June, the Soviets launched two ICBMs, one SLBM and one IRBM while conducting anti-ballistic missile and anti-satellite exercises. One hour after the Soviet ASAT interceptor (Kosmos 1379) was launched from Tyuratam, a Transit-type navigation satellite (Kosmos 1380) was placed in orbit on a C-1 booster from

Plesetsk. This launch, and the subsequent launch of a photo-reconnaissance craft from Tyuratam (Kosmos 1381), marked the first time that any Soviet spacecraft had been orbited in the midst of an ASAT exercise. Yet Soviet mission planners could not have been completely satisfied with their efforts, since Kosmos 1380's final stage malfunctioned, leaving the vehicle in a 140×725-kilometer orbit from which it decayed nine days later.

Second Generation (GLONASS) Soviet Navigation Satellites

Four years after the launch of the American NAVSTAR global positioning system (or GPS) navigation network, the Soviet version (termed GLONASS, for global navigation satellite system) made its debut. Again, the Soviets essentially to have copied the US program. GLONASS vehicles are launched in triplets on a D-1e booster from Tyuratam. The eventual constellation will consist of from nine to twelve spacecraft divided into three orbital planes in semisynchronous 20,000-kilometer orbits, inclined at 65 degrees, with an average orbital period of 12 hours. GLONASS satellites transmit at 1250 and 1603.5 MHz. The large number of satellites in orbit insures that a user will always be within view of three of the vehicles, thus allowing passive-range measurement of ten meters or less. The Soviets have not indicated whether GLONASS is intended for military or civilian use, but its position-location accuracy is so much better than the current military network that it is difficult to imagine it as a purely civilian system. The Soviets may yet follow American practice and make less accurate GLONASS-derived data available to civilian users, perhaps even non-Soviet users.

The first three years of GLONASS operations have revealed growing pains common to many Soviet space efforts. The Soviets have been averaging two triplet launches per year. A full operational constellation has never been established, but enough vehicles have been placed in orbit to facilitate extensive testing (another aspect of GLONASS that closely matches the US GPS program). Until late 1985 all the GLONASS satellites were placed in orbit by means similar to those employed for Soviet geosynchronous launches: a launch to an initial parking orbit, a boost to an elliptical transfer orbit and a final burn to establish the semisynchronous operational orbit. Kosmos 1710–1712, however, entered an initial orbit inclined at 64.8 degrees (GLONASS operational inclination), and then the vehicles were apparently transferred directly to their operational altitude. Western specialists have speculated that this maneuver may be indicative of a new upper stage for the D-1e launch vehicle.

The most disturbing aspect of the GLONASS program, from the Soviet point of view, is its failure rate. Beginning with the first launch, at least one or two members of every triplet payload have failed to achieve a stabilized orbit. This amounts to a 39 percent in-orbit failure rate, which is very high by normal Soviet performance standards, even for a program in its initial phase. Some observers have suggested that the 'dead' GLONASS vehicles are not failures, but rather constitute on-orbit spares or elements that do not have a navigation mission. But the on-orbit spares theory seems unlikely since none of these vehicles has ever been activated, and it is difficult to imagine what other use they could have—particularly in light of recent evidence that these 'dead' vehicles are tumbling in space.

Whatever the true explanation for the large number of inactive second generation Soviet navigation satellites, the USSR's continued series of bi-annual launches indicates a serious commitment eventually to achieve an operational GLONASS network.

Geodetic Soviet Military Satellites

In addition to the low-resolution third generation photo-reconnaissance spacecraft described elsewhere in this volume, the Soviets appear to have deployed at least two other types of geodetic satellites that perform military-related missions. These payloads probably gather data that allows precise determination of geomagnetic anomalies and other information concerning the planet's major land masses.

The first generation of Soviet geodetic military satellites was inaugurated with the launch of Kosmos 203 from Plesetsk on a C-1 booster. Initially, these missions were conducted in near-circular orbits at approximately 1200 kilometers altitude, with an inclination of 74 degrees and an orbital period of around 109 minutes. By the mid-1970s the geodetic mission had changed inclination to 83 degrees. The final first generation launch was conducted in late 1978 (Kosmos 1067). Vehicle weights were probably in the 600 kilogram range.

A month prior to the final first generation mission, the Soviets began testing a new system (Kosmos 1045). A much larger vehicle, perhaps 3400 kilograms in total weight, it required an F-2 booster for orbital insertion. Initially, Western experts classified Kosmos 1045 as an ELINT ferret. The next launch in the series, Kosmos 1312, on 23 September from Plesetsk, was thought to have a communications role. Kosmos 1312 and its successor, Kosmos 1410, were inclined at 83 degrees in near-circular orbits, around 1500 kilometers altitude, with periods of 117 minutes. The placement of Kosmos 1510 into a similar orbit, with an inclination of 74 degrees, in late 1983 seemed to confirm a geodetic mission for all of these spacecraft, given the low frequency of launches and the choice of the first generation geodetic satellite inclinations of either 74 or 83 degrees. Since 1981 the Soviets have established a trend of one launch per year for their second generation military geodetic satellites, usually with inclinations alternating between 74 and 83 degrees.

Anti-satellite Systems

The growing reliance of both superpowers upon space for military purposes has created pressures in both countries to develop and deploy weapons capable of attacking an adversary's space-based assets. Development of anti-satellite (ASAT) technology has proceeded in fits and starts in both the United States and the Soviet Union, usually with pauses occurring in one country while active development is pursued in the other—thus giving the appearance of a leap-frogging arms race. During the early 1960s the Soviets were vocally belligerent about their 'right' to destroy any foreign satellites operating over their territory, but it was the United States that first deployed ASAT missiles in 1963–4 on Kwajalein Atoll and Johnston Island. By the late 1960s the USSR was testing a co-orbital type of interceptor, which, unlike the US nuclear-armed systems, has a conventional 'hot metal' kill capability. Throughout the 1970s rumors circulated in the Western press concerning Soviet development of directed-energy ASAT systems that would be either ground, air or space-based. The United States deactivated its Pacific-based nuclear ASATs in 1974, but by the end of the decade development was proceeding on a new air-launched, conventionally-armed interceptor. After two extensive series of tests, the USSR declared a moratorium on its ASAT program in 1983, while US Department of Defense publications continued to predict deployment of Soviet laser ASATs in the early-to-mid 1990s. The first successful test of the new US ASAT against an operational satellite took place in late 1985. Congress then placed a ban on further tests until October 1986.

ASAT capabilities can be embodied in many types of technology. This section is concerned with those Soviet spaceborne assets that appear dedicated to an ASAT mission. Thus none of the Soviet ground-based ASAT assets (eg, lasers, particle beams or electronic warfare assets that are housed in terrestrial facilities) is examined. Nor are those spacecraft capable of being pressed into improvised ASAT missions considered, since that would include (theoretically at least) any maneuverable Soviet spacecraft currently in operation. Finally, weapon systems with other primary missions (such as ICBMs, SLBMs and ABMs) that could be converted to ASAT roles are also outside the scope of this work.

Co-Orbital interceptor: The most extensive Soviet ASAT program involves the development of a capability to destroy low-orbit enemy satellites. Launched from Tyuratam on an F-1m booster against instrumented target satellites, which are C-1-launched from Plesetsk, the ASAT vehicle conducts an intercept within one or two orbits. A successful 'kill' involves closing to within at least one kilometer of the target vehicle. The ASAT interceptor is then maneuvered away from the target and exploded, apparently in a test of its kill mechanism, which is designed to deploy a cloud of metal pellets in the path of the intended victim, thus destroying the on-coming satellite via kinetic energy. The interceptor can be launched from two pads at Tyuratam, and the Soviets have reportedly stockpiled enough interceptors and associated F-1m launch vehicles to support multiple daily launches.

The Soviets have conducted 20 test interceptions, beginning in 1968. There were two distinct periods of testing, divided by a four-year gap. The first seven tests were conducted between October 1968 and January 1972. The tests, all of which were two-revolution intercepts, achieved a 70 percent success rate. The guidance system employed was evidently an active radar emitter. The second series of Soviet co-orbital ASAT tests commenced in February 1976. The Soviets had evidently been redesigning their interceptor during the four-year hiatus. The most important innovation was a new guidance mechanism, reportedly based on a passive optic-homing sensor, which was first tested in late 1976. This new guidance system would be a major improvement over the old radar sensor, because it would not be susceptible to jamming countermeasures. The new guidance package, however, has yet to be employed in a successful intercept. All six tests conducted with the passive guidance system are regarded as failures by Western experts.

Use of the active guidance package continued in the second series of tests. In seven tests the Soviets could only manage a 57 percent success rate. The majority of successful intercepts has taken place at altitudes

below 1000 kilometers, although the number of successful attacks above this altitude increased in the second series of tests. During the second series of tests the Soviets also successfully demonstrated a single-orbit intercept capability.

The US Department of Defense declared the Soviet ASAT system operational in 1977. The extent of the Soviet threat to Western space assets, however, remains controversial. While the USSR has clearly demonstrated an ability to destroy low-Earth orbit satellites, the interceptor has never been employed against a maneuvering target, it operates within a rather constrained intercept envelope and it employs a guidance system that is vulnerable to jamming. While Soviet ASAT interceptions have never been attempted above an altitude of 1710 kilometers, Western specialists credit the system with an operational ceiling of 5000 kilometers. Table 5 summarizes the history of the Soviet ASAT testing program.

Directed energy and advanced kinetic energy ASAT systems: Rumors persist in Western trade journals that the Soviets are developing orbital ASATs that employ lasers, particle beams or chemically-powered kinetic kill mechanisms. In the late 1970s US government sources quoted in the Western press were predicting a prototype laser ASAT (or LASAT) by the mid-1980s. These predictions have now been advanced to the end of the decade, with operational deployments possible by the mid-1990s.

In the late 1960s the USSR began an R&D effort directed toward spaceborne particle beam generators. The US Department of Defense estimates that the Soviets currently lead the US in the areas of radio-frequency quadrupole accelerators and ion sources for particle beams. The Soviet research efforts could result in a space-based system that uses a particle beam to achieve a 'soft kill' against satellites by disrupting their electronics. A protoype of such a system could be deployed in the mid-1990s. Orbital particle beam weapons that actually destroy their targets are probably not feasible in this century.

The Soviets may be somewhat closer to deploying an advanced orbital kinetic energy ASAT. Alarmist reports circulated in the Western press about the 'actual mission' of Kosmos 1267 when it docked with the Salyut 6 space station in late 1981, but claims that the spacecraft carried multiple launchers for infrared ASAT homing missiles are far from being proven. Kosmos 1267 seems to have been primarily a test of the next generation of the Soviet unmanned cargo craft/space tugs that will be the resupply workhorses for the Mir series of space stations. Yet it is also true that the type of technology claimed for

Kosmos 1267 is now currently being developed in the West for anti-ballistic missile missions, and presumably it could be adapted to the ASAT role even more easily than for the ABM mission. Once the Soviets solve their passive guidance problems for the co-orbital interceptor, this type of orbital ASAT system should also be within their grasp.

Given the shroud of secrecy surrounding all Soviet military space efforts, it is impossible to move beyond speculation concerning the next generation of Soviet ASAT systems. The current Soviet political preference appears to be to try to freeze ASAT technology where it now stands through arms control negotiations before the United States can perfect its miniature homing vehicle interceptor or produce even more exotic ASATs as a result of technology developed under the Strategic Defense Initiative. Should the current Soviet efforts to freeze ASAT technology prove unsuccessful, it would be uncharacteristic if the USSR did not push ahead with development of its own advanced ASAT systems.

Fractional Orbiting Bombardment Systems and Related Launches

In the late 1960s the Soviets evidently developed a fractional orbiting bombardment system (FOBS) capable of delivering strategic weapons against US territory by utilizing Antarctic trajectories. Such a trajectory stands in sharp contrast to the Arctic trajectories employed by ICBMs. The military objective behind FOBS technology would be to reduce significantly the tactical warning time available to a potential opponent. When FOBS entered development in the late 1950s American reliance upon the ground-based radars of the ballistic missile early warning system (BMEWS)—all of which were located in the far north and oriented toward Arctic trajectories—made an attack trajectory over the South Pole strategically attractive. Subsequent US early warning programs (geosynchronous launch detection satellites and phased-array warning radars oriented toward southern approaches) have largely negated the FOBS threat. Such US developments probably explain both the cessation of FOBS testing in 1971 and Soviet willingness to dismantle or convert the 18 FOBS launchers at Tyuratam as part of the SALT II negotiations. Given the failure of the United States to ratify SALT II in the aftermath of the Soviet invasion of Afghanistan, the existence of a residual Soviet FOBS capability remains a matter of some concern in the West.

FOBS testing began in 1967 with Kosmos 139, launched from Tyuratam on an F-1r booster (a derivative of the SS-9 ICBM). Some fifteen tests were conducted over the next four years, the bulk of them in 1967—after which the FOBS was pronounced operational by Western sources. The vehicle weighed around 5000 kilograms and was inserted into fractional orbits with inclinations between 49.5 and 50 degrees. Altitudes varied between 259 to 205 kilometers at apogee, and from 157 to 140 kilometers at perigee. Orbital periods averaged 88 minutes. The typical ground traces for a FOBS test moved northeastward over the Soviet Far East and Siberia; descended toward the equator over the Pacific Ocean; moved on toward the South Pole over South America, ascending across the equator over the Atlantic; and thence went back towards Tyuratam via Africa and the Eastern Mediterranean. The vehicle was then deorbited close to its launch site.

There were also three anomalous F-class vehicle launches that occurred both immediately prior to the FOBS testing program and in its midst. The first two took place in 1966. Both were unannounced (that is, were not even given a Kosmos designator and the usual bland press statement concerning 'scientific apparatus intended for the continuation of research into outer space'). Both missions flew at an inclination of 49.6 degrees in eccentric orbits (1313×156 and 885×140 kilometers, respectively), and both craft were severely fragmented in orbit. The first launch (designated U-1 in the West) was also probably the first F-class vehicle launch.

The third anomaly occurred on 23 December 1969. This time it was announced and was designated Kosmos 316. Its orbit was eccentric (1650×140 kilometers) and the payload was inclined at 49.5 degrees to the equator. This time, however, the vehicle was not fragmented.

There is no satisfactory explanation for these flights. Attempts at explanation in the Western press and specialist literature range from FOBS failures to tests of orbital bomb platforms and experiments involving space mines. The orbits and payload weights differed radically from all other FOBS tests, and the even greater than usual secrecy, employment of F-class boosters and payload fragmentation does in fact strongly suggest some kind of military application. The meaning of these flights remains hidden.

Soviet Minor Military Satellites

'Minor military' is a term used by Western experts to describe a group of small low-orbit Soviet military satellites that are launched from Kapustin Yar and Plesetsk on either the B-1 or C-1 boosters (the last B-1 launch having taken place in June 1977). Some 143 minor military launches have been undertaken between the first recorded launch of such a payload in June 1969 and the end of 1985. Nicholas Johnson of Teledyne Brown Engineering has demonstrated that Soviet minor military satellites can be grouped into two subclasses. The first are placed into eccentric orbits of 82.9 degrees inclination, with apogees ranging from 1600 to 2000 kilometers, perigees between 300 and 400 kilometers and orbital periods falling between 102 and 109 minutes. The second subclass cannot be defined in such detail, but in general it consists of satellites in low-altitude circular orbits, with periods measured between 92 and 95 minutes. The orbits are probably mission-related.

Soviet minor military satellites are thought to perform a variety of missions. The follow-

History of Soviet ASAT Testing

Test Number	Series Number	Date	Kosmos Number	Interceptor Apogee[1]	Perigee[1]	Inclination[2]	Kosmos Number	Target Apogee[1]
1	1	20/10/68	249	1639	502	62.23	248	542
2	1	01/11/68	252	1640	535	62.34	248	543
3	1	23/10/70	374	1053	530	62.96	373	543
4	1	30/10/70	375	994	565	62.86	373	555
5	1	25/02/71	397	1000	575	65.76	394	614
6	1	04/04/71	404	1009	802	65.74	400	1006
7	1	03/12/71	462	1654	231	65.88	459	259
1	2	16/02/76	804	618	561	65.86	803	621
2	2	13/04/76	814	615	556	65.90	803	621
3	2	21/07/76	843	—	—	—	839	2097
4	2	27/12/76	886	1266	532	65.85	880	617
5	2	23/05/77	910	1775	465	65.86	909	2104
6	2	17/06/77	918	1630	245	65.90	909	2106
7	2	26/10/77	961	302	125	65.80	959	834
8	2	21/12/77	970	1148	949	65.85	967	1004
9	2	19/05/78	1009	1362	965	65.87	967	1004
10	2	18/04/80	1174	1025	362	65.83	1171	1010
11	2	02/02/81	1243	1015	296	65.82	1241	1011
12	2	14/03/81	1258	1024	301	65.83	1241	1011
13	2	18/06/82	1379	1019	537	65.84	1375	1012

Source: Nicholas Johnson, *The Soviet Year in Space: 1983* (Colorado Springs, CO: Teledyne Brown, 1984).

ing brief list of possible payload capabilities has been compiled from Western sources.

· groundbased early warning radar calibration
· ABM radar calibration
· ASAT radar calibration
· blue-green communications laser system calibration
· measurement of atmospheric drag
· dispensing of fragments to simulate MIRVed ICBM attack
· advanced materials testing (eg improved ablative shielding for reentry vehicles).

Kosmos (Non-Military)

Historically, a little over 10 percent of Soviet space launches carrying the Kosmos designation have performed non-military missions. As discussed in the previous section, some Kosmos military photo-reconnaissance missions have also carried and/or deployed non-military scientific payloads. In addition, the Kosmos label is frequently used for developmental precursor flights in programs eventually given separate designators (eg Soyuz). Kos-

mos nomenclature also covers both programs that have been discontinued and many failures within on-going separately-designated programs (the latter in an attempt to make these programs appear failure-free).

Table 5 lists the number of non-military Kosmos launches by category. Many of the precursors and failures are discussed under the appropriate program headings elsewhere in this volume. Those missions that do not fall into any other program category are also catalogued in this section.

B-Class flights: The B-1 booster was employed between 1964 and 1973 in a Soviet program that inserted some 36 standardized payloads into orbit from Kapustin Yar and Plesetsk. The satellites weighed between 250 and 425 kilograms, were cylindrical in shape, with hemispheric ends, and came in either battery or solar powered variants. All but two were spin-stabilized. These two interesting weather satellites employed an aerodynamic stabilization system in low-Earth orbit, whereby, deployed rearward from the vehicle, were telescoping booms, at the ends of which an annular ring was attached to provide stability in the thin atmosphere present between altitudes of 250 and 300 kilometers.

Kapustin Yar-launched craft were inclined at 49 degrees until 1966, when the inclination was changed to 48.4 degrees to the equator. Plesetsk launches were given inclinations of either 71 or 82 degrees. A variety of high, medium and low altitude parameters were employed during the nine-year program. These Kosmos missions with the B-1 vehicle were replaced in the mid-1970s by the Interkosmos program, which has relied primarily upon the C-1 booster and is described elsewhere in this volume.

C-Class flights: Since 1970 some eight Kosmos-designated scientific missions have been flown with the C-1 launch vehicle. Two were launched from Kapustin Yar and the rest from Plesetsk. Missions were flown at both circular and eccentric orbits with inclinations of 50.6 (Kapustin Yar), 69.2, 74 and 83 degrees (the latter three inclinations were all associated with the Plesetsk launches). Three types of vehicles appear to have been used in this series:

· a solar-powered modified B-class vehicle weighing around 700 kilograms
· a cylindrical instrumentation package incased in a solar array shell that provided the craft's power; total weight of 320 kilograms

	Perigee[1]	Inclination[2]	Intercept Altitude[1]	Orbits Prior to Intercept	Guidance Package	Mission Outcome
(1)	475	62.25	525	2	Radar	Failure
(2)	473	62.25	535	2	Radar	Success
(3)	473	62.93	530	2	Radar	Failure
(4)	466	62.92	535	2	Radar	Success
(5)	572	65.84	585	2	Radar	Success
(6)	982	65.82	1005	2	Radar	Success
(7)	222	65.83	230	2	Radar	Success
(1)	547	65.85	575	1	Radar	Failure
(2)	549	65.86	590	1	Radar	Success
(3)	983	65.88	1630	2	Radar	Failure[3]
(4)	559	65.85	570	2	Optical	Failure
(5)	993	65.87	1710	1	Radar	Failure
(6)	991	65.87	1575	1	Radar	Success
(7)	144	65.83	150	2	Radar	Success
(8)	963	65.83	995	2	Optical	Failure
(9)	963	65.83	985	2	Optical	Failure
(10)	966	65.85	1000	2	Optical	Failure
(11)	975	65.82	1005	2	Optical	Failure
(12)	976	65.82	1005	2	Radar	Success
(13)	979	65.84	1005	2	Optical	Failure

[1]in kilometers
[2]in degrees
[3]failed to achieve intercept orbit

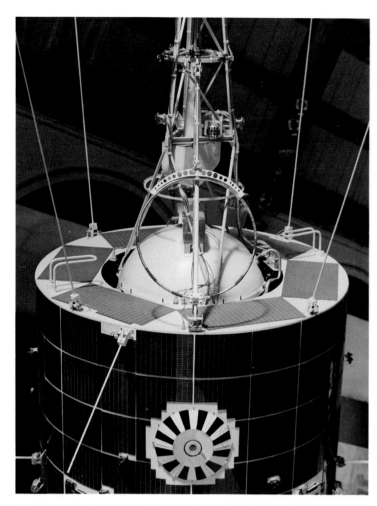

Above: The solar-powered Kosmos 381 scientific satellite. *Below:* The many Kosmos spacecraft led up to (and included some of) the Venera series, *above.*

The International biosat Kosmos 782 carried US livestock. Various national flags encircle the human logo on the apex (*right*) of the space craft. Soviets make use of extant designs—ergo the vessel's Soyuz body. The recovery was 20 days after launch.

a similar cylindrical core, but equipped with eight solar panels that deployed at right angles to the vehicle body once orbit is achieved; total weight was probably around 500 kilograms

The C-class Kosmos scientific launches appear to have terminated in 1977.

A and F-class Kosmos supplemental payloads: Since the initiation of military Kosmos flights in 1962, the Soviets have apparently placed supplemental payloads of a non-military nature upon their military Kosmos vehicles. In addition to taking measurements and performing experiments in the low-Earth orbits employed by the Soviet photo-reconnaissance program, these non-military payloads played a major role in developing weather satellites and in the launch, control and recovery efforts associated with the manned precursor programs. The supplemental payloads were contained in the reentry capsule, affixed to the main vehicle body or deployed at some point during the flight via 'piggyback' mode. Based on data gathered by the Kettering Group and the Royal Aircraft Establishment, the US Congress's Office of Technology Assessment estimates that the USSR launched some 109 non-military supplemental payloads on military Kosmos missions between 1962 and 1980, using A-1, A-2 and F-2 boosters.

Oceanographic research: The Soviets appear to employ the Kosmos label for satellites that advance the study of ocean surface phenomenology. Launched by the F-2 booster, there have been seven such flights since the first (Kosmos 1025) in 1978. From Plesetsk, these spacecraft are inserted into orbits with apogees between 660 and 670 kilometers altitude and perigees between 630 and 650 kilometers altitude, utilizing 82.5-degree inclinations to the equator. Orbital periods averaged between 97 and 98 minutes, and the total vehicle weight was estimated at 6320 kilograms.

The Soviet plan may be to operate the satellites in pairs, with orbital planes spaced 45 degrees apart. Only three of the launches in the series, however, have been described by the Soviets as part of their oceanographic program. Because their orbital parameters closely resemble those of Soviet heavy ELINT satellites, the other four missions of this program have often been classified as ocean ELINT ferrets by Western analysts. The latest launch undertaken in the Soviet oceanographic program was Kosmos 1408 in 1982.

Civilian Kosmos navigation satellites with COSPAS/SARSAT: As mentioned in the military Kosmos section of this volume, the Soviets employ two distinct low-orbit navigation satellite networks. The civilian

network consists of four satellites with orbital planes separated by 45 degrees, inclined at 83 degrees in 1000-kilometer circular orbits, with periods of approximately 105 minutes. These spacecraft supply locational data to Soviet civilian ships and aircraft.

In 1982 the Soviets orbited their first Transit-type navigation satellite equipped with a COSPAS/SARSAT emergency beacon location system (Kosmos 1383). COSPAS is the Soviet acronym for search and rescue satellite (SARSAT in English). A transponder capable of receiving emergency distress signals transmitted at 121.5 and 243 MHz is placed on satellites in nearly polar low-Earth orbits which then relay the location of the distress beacon to ground stations when the vehicles' orbits bring them within line-of-sight. Such a capability greatly reduces the time required to locate and reach ships and airplanes in extremis. The USSR currently has three operational COSPAS receiving sites located at Moscow, Archangel and Vladivostock. A fourth site will probably be built in Siberia.

SARSAT technology was initially developed by the US, USSR, Canada and France. The four founding members have since been joined by the UK, Norway, Bulgaria and Finland, with India soon to participate. The US and the Soviet Union have agreed to maintain at least two active COSPAS/SARSAT-equipped satellites. (The first US vehicle equipped with a SARSAT transponder was NOAA 8, launched in 1983). By mid-1985 SARSAT/COSPAS had been credited with saving at least 500 lives, primarily maritime and aviation accidents.

Kosmos Non-Military Launches by Mission Type

Mission	Number		
Communications	6	Soyuz	11
Earth Resources	3	Soyuz T	5
Geodesy	18	Statsionar	1
Lunar	7	Tug	1
Man-Related	38	Venera	1
Mission Failures		Voskhod	2
Interkosmos	1	Zond	3
Luna	5	Subtotal	44
Mars	5	Scientific	49
Salyut	1	Vehicle Tests	14
Soyuz	1	Weather	15
Venera	9	Total	223
Subtotal	22		
Planetary	7		
Precursor Flights			
Apollo-Soyuz	2		
Lunar Landing (Manned)	4		
Meteor	10		
Molniya	1		
Reentry Vehicle Tests	3		

Sources: US Senate, Committee on Commerce, Science and Transportation, *Soviet Space programs 1976–1980: Part 3* (Washington, DC: US Government Printing Office, 1985): and Nicholas Johnson, *The Soviet Year in Space: 1981, 1982, 1983 and 1985* (Colorado Springs, CO: Teledyne Brown, 1982, 1983, 1984 and 1986).

Soviet Transit-type navigation satellites equipped with COSPAS transponders weigh around 810 kilograms. The transponder appears to remain viable after the navigation mission is passed on to a replacement craft. The Soviets have developed a new 406-MHz beacon that provides much more accurate locational data (to within several kilometers) than the current system. The new beacon was approved in 1985 by the COSPAS/SARSAT steering committee.

Kosmos biosatellites: The Soviets periodically orbit multi-national biological payloads under the Kosmos designation. The last three such flights are described below.

Kosmos 1129
Launch Vehicle: A-2
Launch Site: Plesetsk
Launch Date: 25 September 1979
Recovery Date: 14 October 1979
Total Weight: 5500 kilograms
Apogee: 406 kilometers
Perigee: 226 kilometers
Inclination: 62.8 degrees
Period: 90.5 minutes

This mission tested the reaction of rats to space flight in order to gain further understanding of the effects of zero gravity upon humans.

Kosmos 1514
Launch Vehicle: A-2
Launch Site: Plesetsk
Launch Date: 14 December 1983
Recovery Date: 19 December 1983

Apogee: 259 kilometers
Perigee: 214 kilometers
Inclination: 82.3 degrees
Period: 89.2 minutes

A multi-national effort with participation from Bulgaria, Czechoslovakia, East Germany, France, Hungary, Poland, Rumania, the Soviet Union and the United States, this flight marked the first use of monkeys in the Soviet space program. The monkeys were configured with instruments supplied by the United States to permit study of the circulatory system in zero gravity. Pregnant rats were also aboard the spacecraft so that the embryonic side-effects of zero gravity could be studied. One of the monkeys, named Bion, died shortly after recovery becoming the first primate casualty of the Soviet space program.

Kosmos 1667
Launch Vehicle: A-2
Launch Site: Plesetsk
Launch Date: 10 July 1985
Recovery Date: 17 July 1985
Apogee: 220 kilometers
Perigee: 211 kilometers
Inclination: 82.4 degrees
Period: 89.3 minutes

The latest international biosat had the following cargo:

· two female macaque monkeys named Gordyy and Versiyy
· 10 rats
· 1500 flies

Above: A rather nondescript-looking recovery ship lends additional secrecy to the prototype spaceplane *(foreground)* project labeled 'Kosmos'. See also photos on pages 40 and 84.

· ten newts, each with one foot and an eye lens removed
· Iris flowers

Again, the purpose of the mission was to extract data that would be useful to manned flight applications. Both monkeys survived the flight and initial recovery period. The newts began tissue regeneration in a normal manner when compared to results obtained from similar terrestrial experiments.

Luna

In the early days of the space age the Moon figured prominently in the USSR's efforts to reap as many terrestrial benefits as possible from the early Soviet lead over the United States in large booster capability. In rapid succession Russian spacecraft conducted the first lunar fly-by and lunar strike and returned the first pictures of the dark side of the Moon. In the mid-1960s, the Soviets accomplished the first soft lunar landing and placed the first spacecraft in lunar orbit. All of these spectaculars fired the world's imagination, but they also fueled America's determination to put the first man on the Moon—a feat accomplished by the end of the decade. Unable to compete successfully with the US Apollo program, due both to booster and manned vehicle precursor difficulties (see Appendix II and the Zond entry, respective-

At left: One of two female macaque monkeys aboard Kosmos 1667 is here apparently undergoing an 'interview.' *Above:* Luna 1 was the first attempted Moon probe.

ly), the Soviets concentrated upon automated soil sample-return missions and unmanned lunar rovers (see the Lunokhod entry) during the early-to-mid 1970s. The last Soviet Luna flight took place in 1976, but in 1985 the USSR formally declared its intention to launch a Lunar Polar Orbiter by the end of the decade. Its major mission will be to conduct initial surveys leading to soil sample-return missions from the Moon's dark side. The 24 Luna flights and plans for the circa-1989 Polar Orbiter are described below.

Luna 1

Launch Vehicle: A-1
Launch Site: Tyuratam
Launch Date: 2 January 1959
Total Weight: 361 kilograms
Aphelion: 1318 astronomical units (2)
Perihelion: 0.978 astronomical units (3)
Inclination: 0.01 degrees (4)
Period: 450 days

Luna 1, a spherical vehicle with four protruding antennas and a probe, was the first Soviet attempt to strike the Moon. The vehicle carried a small geophysical package and was emblazoned with the Soviet coat of arms. The spacecraft missed the Moon by around 6000 kilometers and, with the help of lunar gravity, sped out into the solar system as the first artificial satellite to achieve solar orbit.

Luna 2

Launch Vehicle: A-1
Launch Site: Tyuratam
Launch Date: 12 September 1959
Impact Date: 13 September 1959
Total Weight: 390 kilograms
Apogee: 400,000 kilometers

Like its predecessor, Luna 2 was primarily an attempt to impact the Moon's surface via direct ascent trajectory. This attempt was a success, impacting the lunar surface at 1W/30N (about 435 kilometers from the Moon's visible center). The vehicle was essentially identical to Luna 1.

Luna 3

Launch Vehicle: A-1
Launch Site: Tyuratam
Launch Date: 4 October 1959
Decay Date: 20 April 1960
Total Weight: 435 kilograms
Apogee: 476,500 kilometers
Perigee: 40,300 kilometers
Inclination: 73.8 degrees
Period: 15.8 days

The last of the direct ascent missions, Luna 3 was by far the most ambitious. A larger spacecraft than its predecessors, the Automatic Interplanetary Station (as the Soviets called it) was cylindrical, with one hemispheric end and a funnel-shaped end from which sprouted four antennas. Placed in a barycentric orbit, the spacecraft stabilized itself 6200 kilometers from the lunar surface and acquired the first photographs of the dark side of the Moon. These pictures were developed onboard and then transmitted digitally to ground stations when the vehicle's orbit brought it back closer to the Earth. These impressive technical feats were achieved using solar power (Luna 1 and 2 were battery-powered).

Luna (unannounced)
Launch Vehicle: A-2e
Launch Site: Tyuratam
Launch Date: 4 January 1963
Decay Date: 5 January 1963
Total Weight: 1420 kilograms
Apogee: 196 kilometers
Perigee: 167 kilometers
Inclination: 65 degrees
Period: 88.5 minutes

Luna flights resumed following a three-year hiatus using the new A-2e booster (see Appendix II for details), which allowed the Soviets to boost payloads in the 1400 kilogram range from low-Earth orbit to the moon. The first attempt to do so failed, and thus went unacknowledged until the United States revealed the launch as a failure six months later.

Luna 4
Launch Vehicle: A-2e
Launch Site: Tyuratam
Launch Date: 2 April 1963
Total Weight: 1422 kilograms
Initial Apogee: 297 kilometers
Initial Perigee: 167 kilometers
Initial Inclination: 64.7 degrees
Initial Period: 88.2 minutes
Subsequent Apogee: 694,000 kilometers
Subsequent Perigee: 89,250 kilometers
Subsequent Inclination: 65 degrees
Subsequent Period: 29 days

The upper stage functioned correctly on the next Moon shot, but the vehicle missed its target by 8500 kilometers three days after launch and entered a barycentric orbit around the Earth.

Kosmos 60
Launch Vehicle: A-2e
Launch Site: Tyuratam
Launch Date: 12 March 1965
Decay Date: 17 March 1975

Total Weight: 1450 kilograms
Apogee: 287 kilometers
Perigee: 201 kilometers
Inclination: 64.7 degrees
Period: 89.1 minutes

This flight began a year of failures for the Luna program. Kosmos 60 failed to leave low-Earth orbit and thus kept its nondescript label.

Luna 5
Launch Vehicle: A-2e
Launch Site: Tyuratam
Launch Date: 9 May 1965
Impact Date: 12 May 1965
Total Weight: 1476 kilograms
Initial Apogee: 291 kilometers
Initial Perigee: 197 kilometers
Initial Inclination: 65 degrees
Initial Period: 89.3 minutes
Subsequent Apogee: 400,000 kilometers

The second attempt of 1965 did escape Earth orbit, only to suffer a retrorocket failure and crash into the Moon's Sea of Clouds at 31S/8W.

This mission is considered a failure in the West because the Soviets were clearly trying for soft lunar landings at this stage of their moon program, given the much larger payloads than those of the first three Luna flights.

Luna 6
Launch Vehicle: A-2e
Launch Site: Tyuratam
Launch Date: 8 June 1965
Total Weight: 1442 kilograms
Apogee: 246 kilometers
Perigee: 167 kilometers
Inclination: 64.8 degrees
Period: 88.7 minutes

Luna 6 failed because a midcourse correction error caused the vehicle to miss the moon by 160,000 kilometers on 11 June and enter a solar orbit.

Luna 7
Launch Vehicle: A-2e
Launch Site: Tyuratam
Launch Date: 8 June 1965
Impact Date: 7 October 1965
Total Weight: 1506 kilograms
Initial Apogee: 272 kilometers
Initial Perigee: 124 kilometers
Initial Inclination: 64.8 degrees
Initial Period: 88.4 minutes
Subsequent Apogee: 400,000 kilometers

Failure to sustain retrorocket thrust led to Luna 7's destruction on impact at 9N/40W (Sea of Storms).

Luna 8
Launch Vehicle: A-2e
Launch Site: Tyuratam
Launch Date: 3 December 1965
Impact Date: 6 December 1965
Total Weight 1552 kilograms
Initial Apogee: 209 kilometers
Initial Perigee: 169 kilometers
Initial Inclination: 51.8 degrees
Initial Period: 88.2 minutes
Subsequent Apogee: 400,000 kilometers

Above: Luna 1 was the first spacecraft to reach 'second cosmic velocity'—40,234 kph. The launch vehicle was an A-2 Lunick.

Another retrorocket failure caused the last attempt of the year to crash into the Sea of Storms at 9.1N/63.3W.

Luna 9
Launch Vehicle: A-2e
Launch Site: Tyuratam
Launch Date: 31 January 1966

Landing Date: 3 February 1966
Total Weight: 1538 kilograms
Initial Apogee: 219 kilometers
Initial Perigee: 167 kilometers
Initial Inclination: 51.9 degrees
Initial Period: 88.3 minutes
Subsequent Apogee: 400,000 kilometers

In early 1966, the Soviets achieved their goal: the first successful soft lunar landing. In the aftermath of yet another space spec-

tacular the USSR released details of their spacecraft to the West. A modular vehicle, the second generation Luna landers were composed of three main elements. The 100-kilogram payload was contained in a spherical capsule at one end of the essentially cylindrical vehicle. Below the payload was the main body, which contained the flight control instrumentation. Two strap-on modules were attached to either side of the vehicle's waist. These packages held a radar alti-

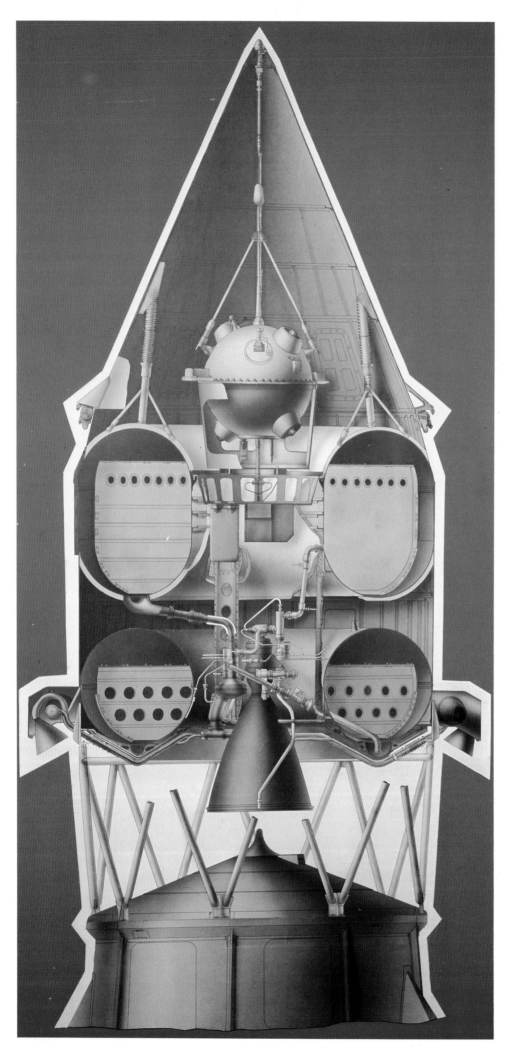

meter, horizontal and vertical astro-orientation sensors and nitrogen micro-thrusters. The vehicle ended in a cone-shaped retrorocket engine, used for midcourse maneuvers and the actual landing. The engine was equipped with a fuel pumping system, fuel tanks and a stabilization system consisting of two vernier nozzles for roll and four vernier nozzles for pitch and yaw.

Descent began with orientation maneuvers 8300 kilometers above the lunar surface. The strap-on modules were cast off less than a minute prior to landing (at about 75 kilometers altitude), and a five-meter ground contact probe was deployed. When the probe touched the lunar surface, the landing engine cut off and the payload (dubbed an Automatic Lunar Station by the Soviets) was lofted from atop the vehicle so as to land separately a short distance away.

Once the ball-shaped capsule, 60 centimeters in diameter, had touched down, using shock absorbers to protect its instruments, the top half opened, and four petal-shaped protective covers were released, thus stabilizing the payload and allowing the deployment and operation of the following equipment:

· a facsimile panoramic television camera
· four brightness standards
· three rod receiving antennas (112 centimeters in height)
· four lobe transmitting antennas (160 centimeters across)

Luna 9 returned the first television pictures from the surface of the Moon. Before the batteries ran out on 6 February, three television sessions (each involving nine positions of the camera's three two-surface mirrors) were conducted.

Kosmos 111
Launch Vehicle: A-2e
Launch Site: Tyuratam
Launch Date: 1 March 1966
Decay Date: 3 March 1966
Total Weight: 1600 kilograms
Apogee: 226 kilometers
Perigee: 191 kilometers
Inclination: 51.9 degrees
Period: 88.6 minutes

This was another failure to leave low-Earth orbit, confirmed by the usual Soviet practice of cloaking unsuccessful missions beneath the Kosmos label. Most Western experts assume that this mission was intended to be a lunar orbiter.

Luna 10
Launch Vehicle: A-2e
Launch Site: Tyuratam

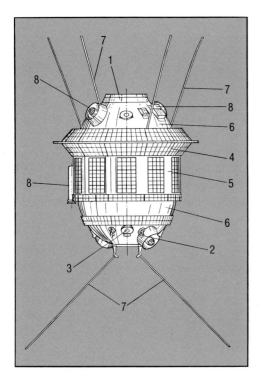

At left: **This cutaway of Luna 1 shows attachment spokes to A-1 booster.** *Above:* **A Diagram of Luna 3: (1) camera window, (2) orientation engine, (3) solar sensor, (4) solar batteries, (5) temperature control shutters, (6) heat shields, (7) aerials, (8) various instruments.** *Right:* **An illustration with data in Russian.**

Launch Date: 31 March 1966
Insertion Date: 3 April 1966
Total Weight: 1600 kilograms
Apogee: 250 kilometers
Perigee: 200 kilometers
Inclination: 52 degrees
Period: 88.5 minutes
Apcynthion: 1017 kilometers (5)
Pericynthion: 350 kilometers
Inclination: 71.9 degrees
Period: 178.3 minutes

In the spring of 1966, the USSR had another space spectacular to announce, the first successful lunar orbit. Luna 10 was a modification of Luna 9, on which the Automatic Lunar Station had been replaced by a 245-kilogram instrument package containing the following types of equipment:

· lunar radiation sensors
· lunar infrared sensors
· solar plasma sensors
· a three-channel magnetometer
· a gamma ray spectrometer
· meteorite particle sensors

The payload was separated from the other modules upon orbital insertion. During its 57-day life Luna 10 made 460 lunar orbits and transmitted 219 broadcasts back to Earth. Experiments included measuring the Moon's magnetic and gravitational fields, determining cosmic ray background radia-

tion and the radiation of lunar rocks and measuring the density of meteorite activity. Luna 10 was even able to regale Earth-bound listeners with the strains of the *Internationale,* thanks to Soviet programming that created certain frequency oscillations in the spacecraft's semiconductors.

Luna 11
Launch Vehicle: A-2e
Launch Site: Tyuratam
Launch Date: 24 August 1966
Insertion Date: 29 August 1966
Total Weight: 1640 kilograms
Apogee: 190 kilometers
Perigee: 177 kilometers
Inclination: 51.9 degrees
Period: 88.1 minutes
Apcynthion: 1200 kilometers
Pericynthion: 160 kilometers
Inclination: 27 degrees
Period: 178 minutes

The Luna 11 batteries lasted 34 days, allowing 277 orbits of the Moon and 137 data transmissions. Subdued Soviet media attention at the time may have indicated that the mission was only partially successful. The Western press reported that the Soviets were unsuccessfully trying to broadcast television pictures from Luna 11. Tass announcements referred to a series of experiments concerning lunar chemical make-up, gravitational fields and meteor activity, as well as various radiation measurements.

Luna 12
Launch Vehicle: A-2e
Launch Site: Tyuratam
Launch Date: 22 October 1966
Insertion Date: 25 October 1966
Total Weight: 1620 kilograms
Apogee: 212 kilometers
Perigee: 199 kilometers
Inclination: 51.9 degrees

transmitting rod antenna than Luna 9 (for a total of four), a radiation sensor mounted next to the camera, an internal dynamograph and four infrared radiometers. It took about 100 minutes to transmit an entire 360-degree view of the spacecraft's surroundings. The batteries probably failed before the new year.

Luna 14
Launch Vehicle: A-2e
Launch Site: Tyuratam
Launch Date: 7 April 1968
Insertion Date: 10 April 1968
Total Weight: 1700 kilograms
Apogee: 242 kilometers
Perigee: 189 kilometers
Inclination: 51.8 degrees
Period: 88.8 minutes
Apcynthion: 870 kilometers
Pericynthion: 160 kilometers
Inclination: 42 degrees
Period: 160 minutes

After a 16-month hiatus, the last second generation Luna spacecraft was launched. The mission appeared similar to Luna 10, and the Soviets subsequently revealed that this flight and Luna 12 carried an electric motor that would be used on the third generation's Lunokhod rover (see separate entry).

Luna 15
Launch Vehicle: D-1e
Launch Site: Tyuratam
Launch Date: 13 July 1969
Impact Date: 21 July 1969
Total Weight: 5600 kilograms
Apogee: 247 kilometers
Perigee: 182 kilometers
Inclination: 51.6 degrees
Period: 88.7 minutes
Initial Apcynthion: 870 kilometers
Initial Pericynthion: 240 kilometers
Inclination: 126 degrees
Period: 160 minutes
Final Apcynthion: 110 kilometers
Final Pericynthion: 16 kilometers
Inclination: 127 degrees
Period: 114 minutes

Like the second generation of the Luna program, the third generation (based on the D-1e launch vehicle, with additional upper stages) got off to a rocky start. The early failures were exacerbated politically by their timing, since they coincided with the culmination of the US Apollo program.

The D-1e launch vehicle allowed the Soviets to place payloads of greater than 5000 kilograms into low-Earth orbit and then to boost them to the Moon. The basic third generation Luna spacecraft was essentially a

Above: Luna 9—upon landing, the black bulb opens to deploy the capsule displayed directly below it, which in turn opens its 'petals' to activate a panoramic TV camera.

Period: 88.6 minutes
Apcynthion: 1200 kilometers
Pericynthion: 133 kilometers
Inclination: 10 degrees
Period: 205 minutes

Luna 12 was a further modification of the basic Luna craft. The instrument package was covered by a large bell-shaped dome. The payload space was probably primarily occupied by television cameras and radio facsimile transmission equipment, which began returning pictures to ground stations on 29 October. Luna 12's maximum resolution was in the range of 15 to 20 meters. The vehicle's batteries lasted for 86 days, allowing 602 orbits and 302 broadcasts.

Luna 13
Launch Vehicle: A-2e
Launch Site: Tyuratam
Launch Date: 21 December 1966

Landing Date: 24 December 1966
Total Weight: 1700 kilograms
Initial Apogee: 223 kilometers
Initial Perigee: 171 kilometers
Initial Inclination: 51.8 degrees
Initial Period: 88.4 minutes
Subsequent Apogee: 400,000 kilometers

The second and last Soviet lunar landing, using the A-2e booster, carried a modified Luna 9 payload. In addition to a panoramic television camera, the capsule was equipped with two mechanical arms that deployed 1.5 meters from the payload sphere. Sensors at the end of the arms included a radiation densitometer, three SBM-10 sensors and a penetrometer. The arms could develop some 23 kilograms of pressure per square meter, in order to collect data concerning lunar soil density and composition. Experimental results revealed a soil depth of 10 to 20 centimeters at Luna 13's landing site (18.9N/ 62W in the Sea of Storms), with a density of around one gram per cubic centimeter.

In addition to the telescoping appendages, the payload spheroid contained one more series of fuel tanks built around two mid-course correction/landing engines. The craft had two pairs of folding landing gear. These were attached to the four large spherical tanks that provided fuel for the descent to the lunar surface. A radar altimeter was also attached to one of the tanks.

Lunar orbital insertion was accomplished using two pairs of cylindrical fuel tanks that were each mounted to a module attached to either side of the descent stage. One insertion module was equipped with electro-optical orientation sensors, and the other held nitrogen microthrusters. Once Lunar orbit was achieved, the insertion stage modules were jettisoned prior to descent in order to decrease weight.

The ascent stage was mounted atop the main vehicle body. It had a single engine supplied by three spherical fuel tanks, above which was mounted an instrumentation and control package. The ascent stage was used to return a spherical reentry capsule to Earth. The reentry capsule sat atop the entire Luna spacecraft.

The D-1e-launched Luna vehicles have been used to perform three basic missions.

1) *Automated Soil Sample Return*: Using the configuration described above, two Luna spacecraft have returned small amounts of lunar soil to Earth for examination.

2) *Lunar Rover Delivery*: By replacing the ascent stage with the Lunokhod vehicle, the heavy Luna craft can be used to ferry the rover from near-Earth orbit to the Moon's surface.

3) *Lunar Orbital Missions*: Using a configuration in which the payload resembles a Lunokhod vehicle without wheels, scientific data is gathered during the life of the heavy Luna vehicles' batteries. For such missions the insertion modules remain attached to the vehicle.

Soviet space planners and mission specialists could not have been pleased with the events of 1969. Rumors persist concerning a failure to launch a Saturn V-type booster in early summer. In mid-July the Soviets inserted Luna 15 into an orbit around the Moon some 48 hours in advance of Apollo 11. The USSR may have been trying to steal some American thunder by returning a lunar soil sample prior to Apollo 11's historic touchdown.

Whatever the Soviet intent, the craft's presence caused anxiety among Apollo 11's mission controllers and appeals were made to Moscow to release Luna 15's orbital parameters. The Soviets did so and gave assurances that the mission would not interfere with Apollo 11. The Soviets sub-

Luna 16, in this illustration, fires its descent engines for touchdown on the Sea of Fertility. The craft's descent stage will be left behind. Earth is seen *upper right*.

sequently changed Luna 15's orbital regime twice. The vehicle attempted a descent on 21 July, while American astronauts Armstrong and Aldrin were on the Moon's surface, but a probable retrorocket failure led to a residual landing speed of 480 kilometers per hour, which destroyed the landing vehicle.

Kosmos 300
Launch Vehicle: D-1e
Launch Site: Tyuratam
Launch Date: 23 September 1969
Decay Date: 27 September 1969
Total Weight: 5600 kilograms
Apogee: 208 kilometers
Perigee: 190 kilometers
Inclination: 51.5 degrees
Period: 88.2 degrees

This was another in the series of anonymous Luna launches which failed to leave low-Earth orbit.

Kosmos 305
Launch Vehicle: D-1e
Launch Site: Tyuratam
Launch Date: 22 October 1969
Decay Date: 22 October 1969

1969 closed with the most dismal failure of the entire Luna program. The main booster apparently malfunctioned, and the payload did not even achieve a single orbit before reentry and destruction.

Luna 16
Launch Vehicle: D-1e
Launch site: Tyuratam
Launch Date: 12 September 1970
Landing date: 20 September 1970
Total Weight: 5600 kilograms
Apogee: 241 kilometers
Perigee: 185 kilometers
Inclination: 51.5 degrees
Period: 88.7 minutes
Initial Apcynthion: 110 kilometers
Initial Pericynthion: 110 kilometers
Inclination: 70 degrees
Period: 119 minutes

Final Apcynthion: 106 kilometers
Final Pericynthion: 15 kilometers
Inclination: 71 degrees
Period: 114 minutes

The first automated lunar soil sample-return mission touched down in the Moon's Sea of Fertility at 0.4N/56.1E. The sample was collected by means of an extendable appendage with a drill attached to the end. The extracted soil was placed in a container that was sent mechanically to the top of the spacecraft and inserted into the reentry capsule. The Soviets ceased drilling at a depth of 35 centimeters when the drill encountered rock thought solid enough to damage the mechanism.

The ascent stage (weighing perhaps 470 kilograms) lifted off on 21 September, and, three days later, the reentry capsule landed in the USSR carrying some 101 grams of lunar soil. The ascent stage followed a ballistic trajectory for the return flight.

Luna 17

Launch Vehicle: D-1e
Launch Site: Tyuratam
Launch Date: 10 November 1970
Landing Date: 17 November 1970
Total Weight: 5600 kilograms
Apogee: 237 kilometers
Perigee: 202 kilometers
Inclination: 51.5 degrees
Period: 88.7 minutes
Initial Apcynthion: 85 kilometers
Initial Pericynthion: 85 kilometers
Inclination: 141 degrees
Period: 116 minutes
Final Apcynthion: 85 kilometers
Final Pericynthion: 19 kilometers
Inclination: 141 degrees
Period: 114 minutes

Luna 17 landed in the Sea of Rains and deployed Lunokhod 1, the first unmanned lunar rover. Lunokhod 1 reached the surface on a pair of folding ramps that were placed, along with the rover itself, on top of the descent stage, in place of the ascent module. Lunokhod functioned for almost a year, with its mission terminating officially on the 24th anniversary of the first Sputnik flight. The Lunokhod vehicle and the details of its two missions are discussed in the next section of this volume.

Luna 18

Launch Vehicle: D-1e
Launch Site: Tyuratam
Launch Date: 2 September 1971
Landing Date: 11 September 1971
Total Weight: 5600 kilograms
Apogee: 242 kilometers
Perigee: 186 kilometers

Inclination: 51.6 degrees
Period: 88.7 minutes
Initial Apcynthion: 101 kilometers
Initial Pericynthion: 101 kilometers
Inclination: 51.6 degrees
Period: 119 minutes
Final Apcynthion: 100 kilometers
Final Pericynthion: 18 kilometers
Inclination: 35 degrees
Period: 114 minutes

Luna 18 was probably an attempted soil sample-return mission. The spacecraft either impacted at too fast a residual speed and was destroyed, or, as the Soviets were to claim, the terrain was too rough even with a soft landing. In any event, the spacecraft stopped returning signals at touchdown.

Luna 19

Launch Vehicle: D-1e
Launch Site: Tyuratam
Launch Date: 28 September 1971
Insertion Date: 3 October 1971
Total Weight: 5600 kilograms
Apogee: 260 kilometers
Perigee: 172 kilometers
Inclination: 51.6 degrees
Period: 88.8 minutes
Initial Apcynthion: 140 kilometers
Initial Pericynthion: 140 kilometers
Inclination: 40.6 degrees
Period: 121.8 minutes
Subsequent Apcynthion: 385 kilometers
Subsequent Pericynthion: 77 kilometers
Inclination: 40.6 degrees
Period: 131 minutes

The first third generation Luna orbital mission was launched less than a month after the ill-fated Luna 18. The scientific payload conducted experiments that involved measuring lunar magnetic fields, cosmic and solar radiation and meteorite activity.

Observations of solar activity were compared with data collected by similar instruments on Mars 2 and 3, Prognoz 1 and 2 and Venera 7 and 8.

Luna 20

Launch Vehicle: D-1e
Launch Site: Tyuratam
Launch Date: 14 February 1972
Landing Date: 21 February 1972
Total Weight: 5600 kilograms
Apogee: 238 kilometers
Perigee: 191 kilometers
Inclination: 1.5 degrees
Period: 88.7 minutes
Initial Apcynthion: 100 kilometers
Initial Pericynthion: 100 kilometers
Inclination: 65 degrees
Period: 118 minutes
Final Apcynthion: 100 kilometers
Final Pericynthion: 21 kilometers
Inclination: 65 degrees
Period: 114 minutes

The second automated lunar soil sample-return mission touched down in the Sea of Fertility near the site of the Luna 18 failure, giving rise to even more questions concerning the fate of the 1971 landing attempt. If, as Moscow had claimed, the loss of Luna 18's signal at touchdown was due to rough terrain, why send the next lander into the same area? Luna 20 seems to have had no problems with the terrain, although it collected a sample only about half the size of that returned by Luna 16's.

The ascent stage lifted on 22 February, and the reentry capsule was recovered about three days later. The soil, which had been taken from the lunar highlands, was found to be lighter and denser than the previous sample. Analysis revealed 70 chemical trace elements in the sample.

Luna 21

Launch Vehicle: D-1e
Launch Site: Tyuratam
Launch Date: 8 January 1973
Landing Date: 16 January 1973

At left is Luna 16, displayed with charts of its trajectory. *Above:* The Moon in the trees: Luna 20's soil sample returns to Earth. *At right* is a portrait of Luna 17.

Total Weight: 5600 kilograms
Apogee: 236 kilometers
Perigee: 183 kilometers
Inclination: 51.6 degrees
Period: 88.6 minutes
Initial Apcynthion: 110 kilometers
Initial Pericynthion: 90 kilometers
Inclination: 60 degrees
Period: 118 minutes
Final Apcynthion: 110 kilometers
Final Pericynthion: 16 kilometers
Inclination: 60 degrees
Period: 114 minutes

Luna 21 delivered the Lunokhod 2 lunar rover to a landing site some 180 kilometers north of that of Apollo 17 in the Le Monier Crater (Sea of Serenity). The rover had several improvements over the initial model. (These are discussed in the next section.) In general, Lunokhod 2 covered four times the territory in half the time of its predecessor before shutting down in early June.

Luna 22

Launch Vehicle: D-1e
Launch Site: Tyuratam
Launch Date: 29 May 1974
Insertion Date: 2 June 1974
Total Weight: 5600 kilograms
Apogee: 227 kilometers
Perigee: 178 kilometers
Inclination: 51.6 degrees
Period: 88.5 minutes
Initial Apcynthion: 221 kilometers
Initial Pericynthion: 219 kilometers
Inclination: 14.6 degrees
Period: 130 minutes

Subsequent Apcynthion: 244 kilometers
Subsequent Pericynthion: 25 kilometers
Inclination: 14.6 degrees
Period: 121 minutes
Final Apcynthion: 1578 kilometers
Final Pericynthion: 30 kilometers
Inclination: 21 degrees
Period: 142 minutes

This lunar orbital mission continued to collect data sought by Luna 19. Lunar plasma, gravitational and magnetic fields were measured; solar and meteorite activity were examined; and large tracts of lunar terrain were acquired on high-resolution photography. Luna 22's reserve of maneuvering fuel lasted 15 months, allowing 30,000 radio commands from Soviet mission controllers, 2400 data broadcasts from the spacecraft to Earth and 1500 trajectory measurements.

Luna 23

Launch Vehicle: D-1e
Launch Site: Tyuratam
Launch Date: 28 October 1974
Landing Date: 6 November 1974
Total Weight: 5600 kilograms
Apogee: 246 kilometers
Perigee: 183 kilometers
Inclination: 51.5 degrees
Period: 88.7 minutes
Initial Apcynthion: 104 kilometers
Initial Pericynthion: 94 kilometers
Inclination: 138 degrees
Period: 117 minutes
Final Apcynthion: 105 kilometers
Final Pericynthion: 11 kilometers
Inclination: 138 degrees
Period: 114 minutes

Luna 23's landing at 13.5N/56.5E in the Sea of Crises was successful, but the rugged terrain apparently caused a malfunction of the drilling rig. The truncated mission was terminated seven days after touchdown, and no sample return was attempted.

Luna 24

Launch Vehicle: D-1e
Launch Site: Tyuratam
Launch Date: 9 August 1976
Landing Date: 18 August 1976
Total weight: 5600 kilograms
Apogee: 242 kilometers
Perigee: 188 kilometers
Inclination: 51.5 degrees
Period: 88.7 minutes
Initial Apcynthion: 115 kilometers
Initial Pericynthion: 115 kilometers
Inclination: 120 degrees
Period: 119 minutes
Final Apcynthion: 120 kilometers
Final Pericynthion: 12 kilometers
Inclination: 120 degrees
Period: 114 minutes

The almost two-year hiatus in Luna flights that followed the Luna 23 failure was probably used to redesign the drilling rig for what turned out to be both the last soil sample return and the last Soviet lunar landing to date. Luna 24 touched down near the Luna 23 landing site at 12.7N/62.2 E. The mission goal was to extract a core sample of around 2.5 meters in depth. Both previous successful efforts had only managed to reach a depth of 30 centimeters, primarily due to encounters with large rocks.

The improved drilling rig was equipped with an impact-rotating device that changed drilling speed depending upon the type of material resistance encountered, thus allowing the drill to penetrate dense rock. Sample integrity was assured, and drilling tube blockage was avoided via an externally-fed plastic liner for the tube that wrapped around the core sample as the drill moved deeper below the lunar surface. A winch mounted next to the reentry capsule then extracted the plastic-wrapped core sample from the drilling tube and coiled it around a drum. The drum was inserted into the return capsule, and the drilling rig swung away from the ascent stage to facilitate lift-off.

The ascent stage was launched on 19 August and was recovered three days later in the USSR. The total sample weight was 170 grams. According to *Pravda*, mass spectros-copy and x-ray microanalysis revealed traces of 60 chemical elements in the lunar soil sample.

The Lunar Polar Orbiter

In March 1985 The Soviet Union announced official intention to proceed with the Lunar Polar Orbiter program by the turn of the decade. The spacecraft will probably be a variant of the heavy Luna vehicle and will be placed into a 100-kilometer lunar polar orbit. Two primary mission objectives have been announced: mapping and geophysical experiments. The mapping function will be carried out with a television system of unspecified resolution. One of the major purposes of this mission is to locate future landing sites (especially on the Moon's dark side) for a resumption of soil sample return flights.

The 300-kilogram instrumentation package will also contain the following equipment designed to continue research efforts initiated by Luna 19 and 22:

· six spectrometers (germanium gamma, scintillation gamma, x-ray, charged particle, reflected electron and infrared)

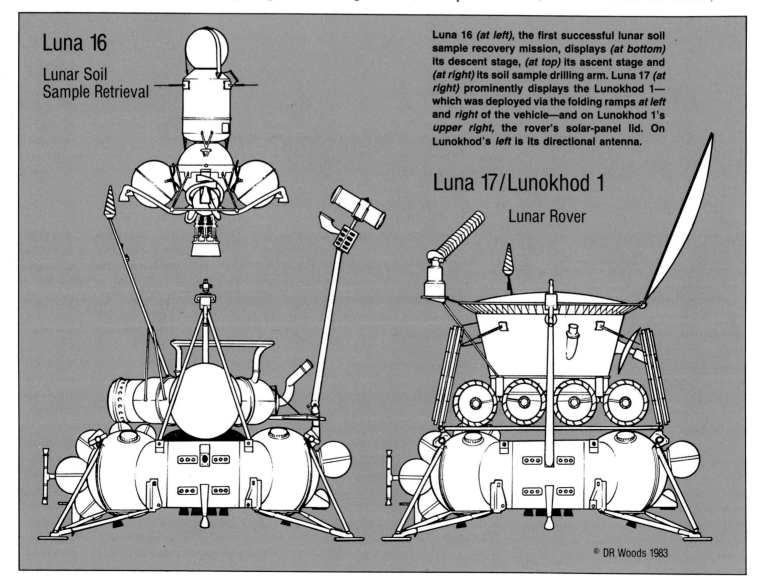

Luna 16

Lunar Soil
Sample Retrieval

Luna 16 *(at left),* the first successful lunar soil sample recovery mission, displays *(at bottom)* its descent stage, *(at top)* its ascent stage and *(at right)* its soil sample drilling arm. Luna 17 *(at right)* prominently displays the Lunokhod 1—which was deployed via the folding ramps *at left* and *right* of the vehicle—and on Lunokhod 1's *upper right,* the rover's solar-panel lid. On Lunokhod's *left* is its directional antenna.

Luna 17/Lunokhod 1

Lunar Rover

© DR Woods 1983

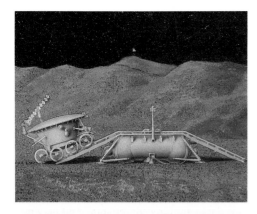

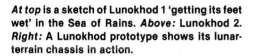

At top is a sketch of Lunokhod 1 'getting its feet wet' in the Sea of Rains. *Above:* Lunokhod 2. *Right:* A Lunokhod prototype shows its lunar-terrain chassis in action.

· three detectors (micrometeorite, neutron, and plasma)
· a spectrophotometer
· a magnetometer
· a radar altimeter

Lunokhod

Lunokhod 1: In late 1970, the Soviets landed the first of two unmanned lunar rovers. Lunokhod resembled a mushroom on wheels (actually, two sets of four independently-powered wheels mounted on either side of its 'stem'). The vehicle was solar-powered, with the cells mounted on the underside of a hinged lid on the top of the rover (or, to continue the mushroom analogy, the top of the 'cap' swung open to reveal the solar panels). Thus Lunokhod 1 was forced to shut down during the two-week-long lunar nights. During periods of intense cold caused by the lunar darkness, the vehicle's equipment was kept functional by an internal heating system powered by a radioisotope source.

Lunokhod was maneuvered by a four-member control team at a ground station on Earth. Four television cameras provided 360 degrees of vision. Directions were received from, and data transmitted back to, Earth via a cone-shaped antenna and a helical antenna mounted near the top of the vehicle. Lunokhod 1 had two speeds forward and backward, as well as a specially designed suspension system for lunar terrain. Fail-safe systems were installed, whereby movement would cease if sensors detected too great a degree of lateral tilt or too steep a slope.

Lunokhod 1 weighed 756 kilograms and carried a penetrometer for determining soil density. Chemical soil tests were accomplished with an x-ray spectrometer mounted on the side of the vehicle base opposite from the penetrometer assembly. The four cameras provided close-up, panoramic and stereo views. Other scientific instrumentation included cosmic ray detectors and a French-built laser detector that reflected laser beams originating from Earth, allowing very precise measurements of both lunar and terrestrial objects.

Lunokhod 1 exceeded its design life by a factor of four, operating around 12 lunar days (almost 365 terrestrial days). The vehi-cle travelled a total of over ten and a half kilometers and took 20,000 television pictures, including 206 panoramic views. Soil experiments included some 500 mechanical tests and 25 chemical tests.

Lunokhod 2: The second rover featured many improvements to the original design. The vehicle weighed 84 kilograms more than its predecessor. Its nuclear heater was powered by Polonium 210. Lunokhod 2 could travel twice as fast as the original rover, an improvement which necessitated the use of a five-member control crew. Other new equipment included:

· an additional camera mounted to provide a better view of terrain ahead of the vehicle
· ultra-violet sensors
· an astrophotometer
· a pole-mounted magnetometer that extended some 2.5 meters ahead of the rover

In all, Lunokhod 2 traveled 35 kilometers from the eastern edge of the Sea of Serenity to the Taurus mountains. Some 80,000 television pictures were returned to Earth, including 86 panoramic views. Lunokhod 2 conducted 740 mechanical soil tests. About

4000 laser contacts were made during lunar darkness by the French-built 14-element retroreflector array. This instrument allowed French and Soviet scientists to discern shifts of the Earth's poles as small as 10 centimeters and to measure the distance between the Earth and the Moon to within 30 centimeters.

Mars

The contrast between the Soviet Mars and Venera programs could not be more stark. While the Soviets have achieved considerable success in their efforts to place probes in orbit around, and on the surface of, Venus, their efforts with respect to Mars can only be described as dismal. Only one of the seven Mars flights was a complete success. The seven missions given the Mars designation are described below, as well as those Kosmos-named failures that, had they at least left Earth-orbit, would probably have been given a place in the Mars

nomenclature. In addition, this section examines the early unannounced Mars failures that were later revealed by the US government. Those flights to Mars that were attempted with Zond spacecraft are discussed in the section devoted to that particular Soviet designator.

Unannounced Mars Attempts

The Soviets appear to have attempted missions to Mars during every launch window opportunity from 1960 through 1973. In an effort to derive maximum propaganda value from what would have been man's first attempt to send spacecraft to the Red Planet, Nikita Krushchev timed his arrival in New York for the 1960 United Nations' opening session to coincide with two A-2e launches from Tyuratam on 10 and 14 October. The Soviet leader reportedly carried models of the spacecraft in his luggage for display to the Western media. Neither launch, however, managed to reach a near-Earth orbit from which it could be lofted toward Mars. Krush-

chev's models remained packed, and the nature of the attempted missions remained unknown until revealed by the United States until 1962.

For the 1962 launch window the Soviets orbited three payloads, all with the A-2e booster. The one launched on 24 October reached an initial orbit of 485x180 kilometers, with an inclination of 64.9 degrees and a period of 91.2 minutes. It failed to launch toward Mars and decayed on 29 October. The spacecraft launched on 1 November was Mars 1 and is described below. The payload on 4 November achieved an orbit of 590x197 kilometers, inclined at 64.7 degrees to the equator, with a period of 92.4 minutes, in which it was stranded and decayed on 19 January 1963.

There may have been additional unannounced Mars attempts in 1964 and 1968. On 30 November 1964, Zond 2 was launched toward Mars. The Soviet practice of attempting multiple launches during each Mars window makes it likely that another

Mars mission failed to achieve orbit during this time. The 1969 window came and went without any Mars, Zond or Kosmos craft attaining orbit, but there were persistent rumors in the Western press concerning Mars-related launch failures that did not reach orbit.

Announced Attempts

Mars 1
Launch Vehicle: A-2e
Launch Site: Tyuratam
Launch Date: 1 November 1962
Fly-By Date: 19 June 1963
Apogee: 238 kilometers
Perigee: 157 kilometers
Inclination: 65 degrees
Period: 88.4 minutes
Aphelion: 1.604 astronomical units
Perihelion: 0.924 astronomical units
Inclination: 2.68 degrees
Period: 519 days

Although successfully launched from Earth orbit, Mars 1 stopped communicating with ground stations on 21 March 1963. The closest approach to Mars was 193,000 kilometers on 19 June 1963, after which the spacecraft entered a heliocentric orbit. The vehicle itself was of the modified Venera-type that became the basic Soviet platform for solar system exploration.

Kosmos 419
Launch Vehicle: D-1e
Launch Site: Tyuratam
Launch Date: 10 May 1971
Decay Date: 12 May 1971
Total Weight: 4650 kilograms
Apogee: 174 kilometers
Perigee: 159 kilometers
Inclination: 51.4 degrees
Period: 88.7 minutes

While Kosmos 419 failed to leave low-Earth orbit (thus retaining its generic designator), the flight did mark the first use of the D-1e launch vehicle for a planetary mission. The larger booster afforded a heavier payload, which was required for the series of orbiter/lander missions launched during the 1971 window.

Mars 2
Launch Vehicle: D-1e
Launch Site: Tyuratam
Launch Date: 19 May 1971
Insertion Date: 27 November 1971
Landing Date: 27 November 1971
Total Weight: 4650 kilograms
Apogee: 173 kilometers
Perigee: 137 kilometers
Inclination: 51.5 degrees

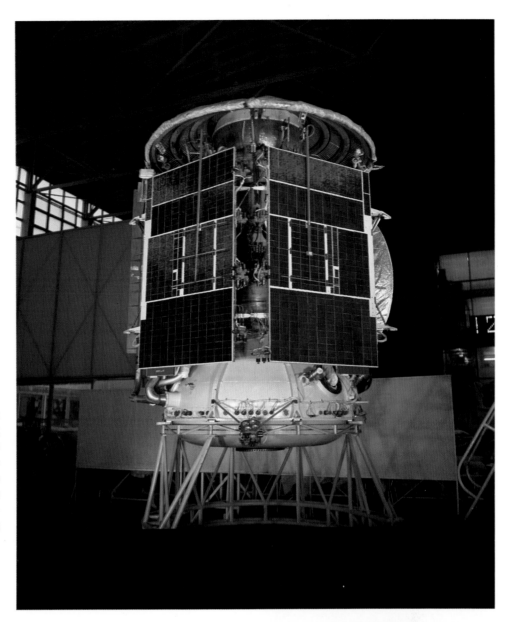

Period: 87.5 minutes
Aphelion: 1.57 astronomical units
Perihelion: 0.99 astronomical units
Inclination: 2.2 degrees
Period: 530 days
Apares: 25,000 kilometers (6)
Periares: 1380 kilometers
Inclination: 18.9 degrees
Period: 1098 minutes

Mars 2 carried both a package of instruments for orbital geophysical measurements and a 635-kilogram lander. The orbiter carried cameras with wide-angle and narrow-angle lenses. The system exposed 12 frames of film simultaneously, developed them on board and then transmitted them back to Earth in digital format. Other instruments included in the geophysical payload were for measuring Martian surface temperature via infrared radiometry, discerning water vapor concentrations via spectroscopy, measuring surface relief with an infrared spectrometer and measuring the density of the upper Martian atmosphere via ultraviolet photometry.

Left: Lunokhod 1 conducted RIFMA tests on lunar rocks. *At top:* This view of Mars 3, which met its end in a Martian dust storm, is the complement to the Mars 3 view *above.*

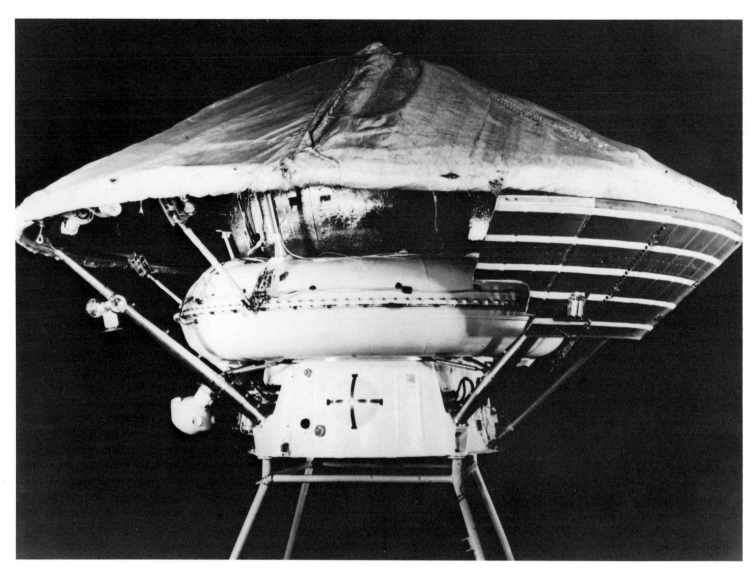

At right: Mars 3 is atop its orbital apparatus, the bulk of which is a fuel tank. Trombone-like tube near the right 'wing' is a thermal radiator. *Above:* The Mars 3 Reentry Module.

The lander was provided with a panoramic television system. It also had instruments to measure atmospheric properties (temperature, pressure and wind velocity sensors, as well as a mass spectrometer) and equipment to conduct chemical and mechanical tests on soil samples. An omnidirectional antenna transmitted data from the lander to the orbiter, which then relayed the information to Earth via high gain antennas.

The lander was deorbited on 27 November 1971 and touched down at 44S/47E. The Soviets announced this event three days later, but gave no further information. Presumably the vehicle failed at or before touchdown. The Mars 2 orbiter continued to return data at least through March 1972. The Soviets officially terminated the Mars 2 and 3 missions on 22 August.

Mars 3
Launch Vehicle: D-1e
Launch Site: Tyuratam
Launch Date: 28 May 1971
Insertion Date: 2 December 1971
Landing Date: 2 December 1971
Total Weight: 4650 kilograms
Apogee: 234 kilometers
Perigee: 140 kilometers
Inclination: 51.6 degrees
Period: 88.2 minutes
Aphelion: 1.57 astronomical units
Perihelion: 0.99 astronomical units
Inclination: 2.2 degrees
Period: 530 days
Apares: 190,700 kilometers
Periares: 1500 kilometers
Inclination: 18.9 degrees
Period: 11 days

The Mars 3 spacecraft was identical to the Mars 2 lander/orbiter combination. The Mars 3 lander deorbited and touched down at 45S/158W on 2 December. Signals ceased after only 20 seconds of transmissions from the surface, probably due to dust storms.

The two orbiters, however, returned a significant amount of data between December and March. Atomic hydrogen and oxygen were detected in the upper Martian atmosphere, which was determined to have a density of .1 of that of Earth. A temperature map of the Martian surface was charted, as well as parameters concerning terrain relief. Atmospheric content and water vapor densities were determined, and the Martian magnetic field was measured.

Mars 4
Launch Vehicle: D-1e
Launch Site: Tyuratam
Launch Date: 21 July 1973
Fly-By Date: 10 February 1974
Total Weight: 3440 kilograms
Apogee: 179 kilometers
Perigee: 147 kilometers
Inclination: 51.5 degrees
Period: 87.5 minutes
Aphelion: 1.63 astronomical units
Perihelion: 1.02 astronomical units
Inclination: 2.2 degrees
Period: 556 days

The 1973 launch window was less favorable to the Soviets than the windows had been for their previous Mars missions. This was due to the relative positioning of the Earth and Mars, which necessitated a higher-energy trajectory to reach the Red Planet. The Soviets chose to overcome this obstacle by splitting the components of the orbiter/lander vehicles, sending two orbiters and

two landers each on its own dedicated vehicle. Mars 4 was an orbiter that reached Mars in early February 1974 but could not achieve orbital insertion, due to a retrorocket malfunction. The vehicle made a close fly-by (2200 kilometers) of the planet, and returned photographs via facsimile scan before entering a heliocentric orbit.

Mars 5
Launch Vehicle: D-1e
Launch Site: Tyuratam
Launch Date: 25 July 1973
Insertion Date: 2 February 1974
Total Weight: 3440 kilometers
Apogee: 174 kilometers
Perigee: 159 kilometers
Inclination: 51.6 degrees
Period: 87.8 minutes
Aphelion: 1.65 astronomical units
Perihelion: 1.01 astronomical units
Inclination: 2.2 degrees
Period: 567 days
Apares: 32,500 kilometers
Periares: 1760 kilometers
Inclination: 35 degrees
Period: 1493 minutes

Mars 5, also an orbiter, was essentially identical to Mars 4. This vehicle's performance constituted the only complete success associated with the 1973-74 Mars missions. The following instruments were used to conduct geophysical experiments after the spacecraft entered an elliptical Martian orbit on 2 February:

· a two-camera television system equipped with red, blue, green and orange filters
· a radio telescope
· a radio probe
· six photometers
· two polarimeters
· an infrared radiometer
· a spectrometer

Experiments included measurements of water vapor and ozone density in the upper Martian atmosphere. The photographs taken by Mars 4 and 5 were eventually combined with those acquired by the US Mariner 9 probe in order to produce surface maps of Mars. French equipment for studying solar radio emissions and proton and electron fluxes was carried on the four Mars spacecraft launched in 1973.

Mars 6
Launch Vehicle: D-1e
Launch Site: Tyuratam
Launch Date: 5 August 1973
Landing Date: 12 March 1974

Apogee: 193 kilometers
Perigee: 154 kilometers
Inclination: 51.5 degrees
Period: 87.9 minutes
Aphelion: 1.67 astronomical units
Perihelion: 1.01 astronomical units
Inclination: 2.2 degrees
Period: 567 days

Mars 6 carried the first of the landers launched during the 1973 window. It reached Mars in mid-March of 1974 and the 1200-kilogram lander was successfully de-orbited, touching down at 24S/25W. Data ceased returning from the lander, however, 148 seconds after its parachute opened.

Mars 7
Launch Vehicle: D-1e
Launch Site: Tyuratam
Launch date: 9 August 1973
Fly-By Date: 9 March 1974
Total weight: 3260 kilograms
Apogee: 193 kilometers
Perigee: 154 kilometers
Inclination: 51.5 degrees
Period: 87.9 minutes
Aphelion: 1.69 astronomical units
Perihelion: 1.01 astronomical units
Inclination: 2.2 degrees
Period: 574 days

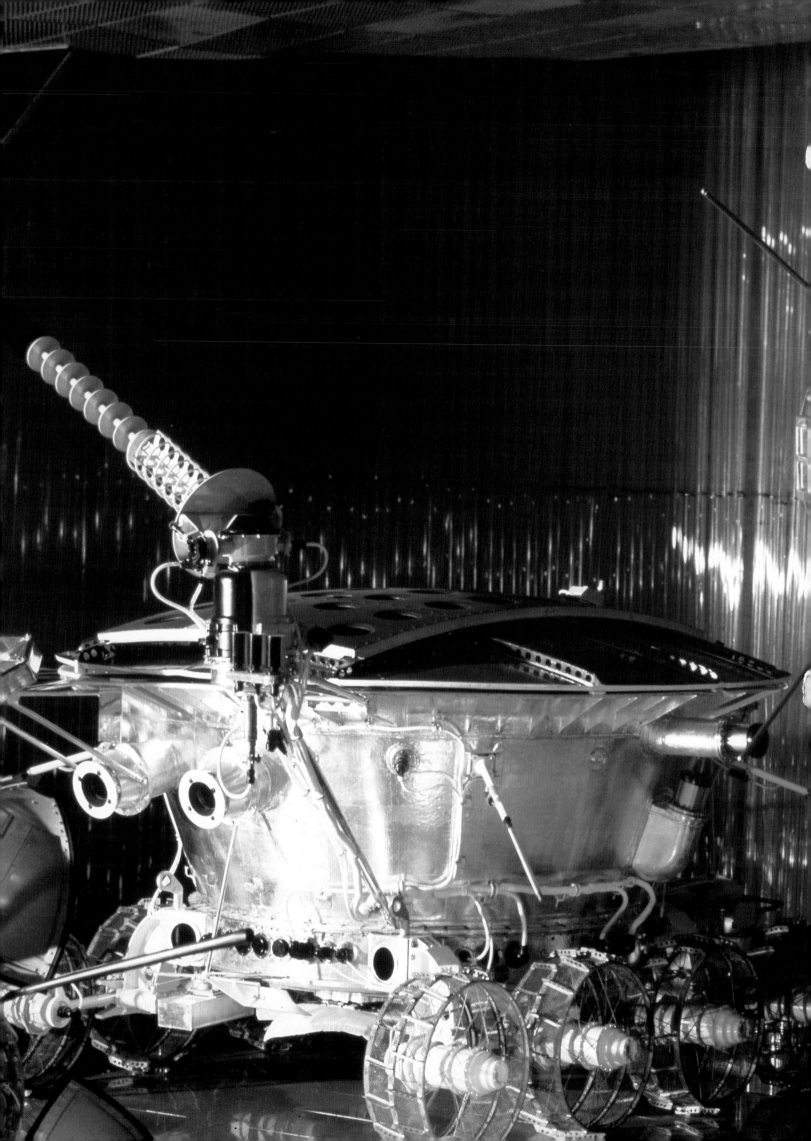

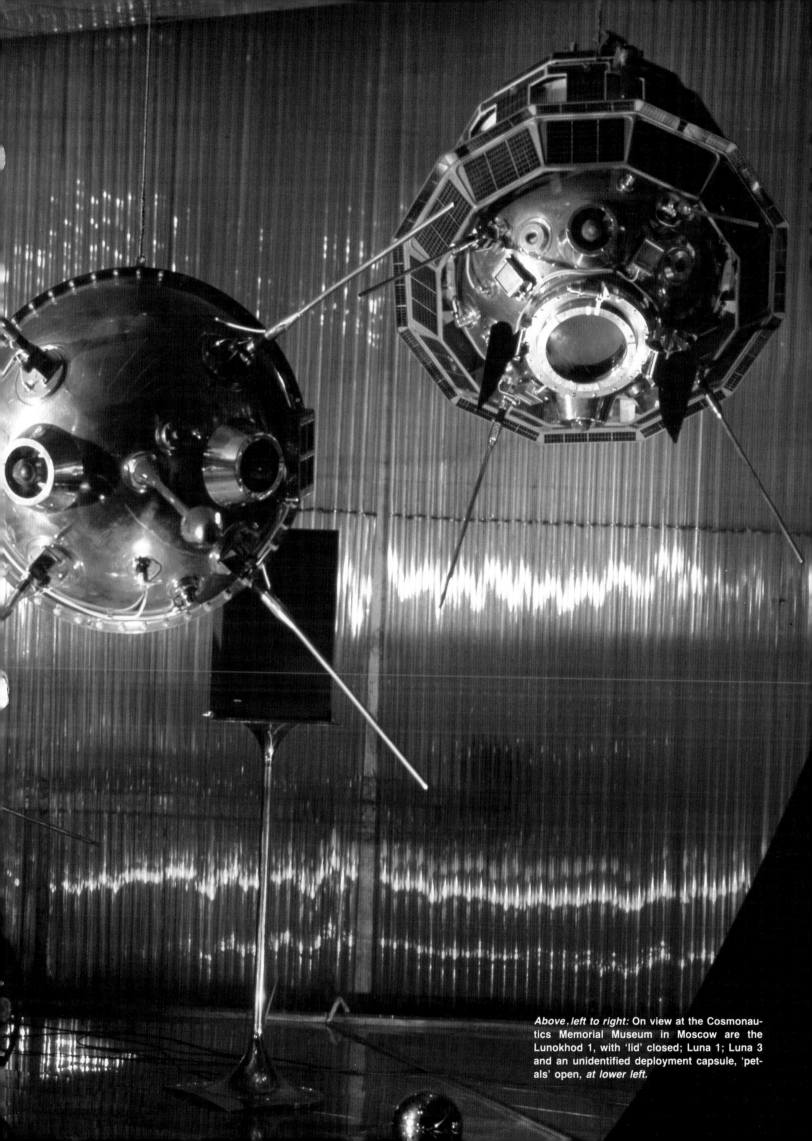

Above, left to right: On view at the Cosmonautics Memorial Museum in Moscow are the Lunokhod 1, with 'lid' closed; Luna 1; Luna 3 and an unidentified deployment capsule, 'petals' open, *at lower left.*

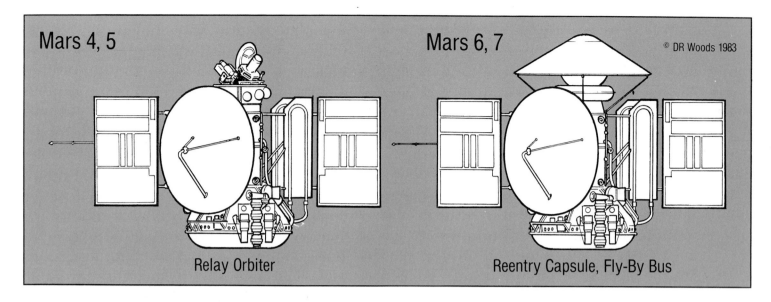

Mars 4, 5 — Relay Orbiter

Mars 6, 7 — Reentry Capsule, Fly-By Bus

© DR Woods 1983

The second lander also reached Mars in March, but a retrorocket failure led to the spacecraft's missing Mars by 1300 kilometers. The lander had been aimed at 50S/28W on the planet's surface.

Phobos 88

The USSR is planning to restart its Mars exploration program, after a decade during which no launches toward Mars were undertaken, with a major international effort directed toward the Martian moon Phobos. The project calls for two spacecraft based on Venera follow-on technology to be launched in July 1988. A series of different orbits would eventually bring one of the vehicles within 50 meters of the surface of Phobos, at which point a lander would be placed on the planetoid's surface. Two candidate lander designs are currently under development: a long-term stationary type and a short-term 'hopper'. The long-term lander would operate for 12 months, conducting a series of seismic, soil, temperature, geophysical and celestial mechanics experiments. The hopper would be equipped with an extendable support that would propel it for a distance of approximately 20 meters in the low gravity of the Martian moon. The hopper would employ a gravimeter, a dynamograph, an x-ray florescence spectrometer and a dynamograph for experiments between hopping maneuvers. In all, the hopper would probably make some 10 hopping maneuvers over a period of a month before expending its batteries. The Soviets are keeping their options open as to which of the two landers, or combination of both, will be sent to Phobos. They are also debating whether an attempt will also be made during the mission to place a lander on the other Martian moon, Deimos. A dozen nations and organizations have made commitments to this project. The Soviets are also working on plans for a Martian rover, probably based on Lunokhod technology, for a resumption of the Martian soft-landing program, which has been dormant since 1973.

Mas

The MAS flights were part of the continuing Franco-Soviet cooperative space effort, which, with the possible exception of the Apollo-Soyuz Test Program, is the most extensive cooperative space effort involving the USSR and a Western industrial democracy. The two MAS missions were small piggyback payloads added to Molniya communications satellite launches. Their orbital elements and mission details are discussed below.

Mas 1

Launch Vehicle: A-2e
Launch Site: Plesetsk
Launch Date: 4 April 1972
Decay Date: 26 February 1974
Total Weight: 15 kilograms
Apogee: 39,250 kilometers
Perigee: 458 kilometers
Inclination: 65.6 degrees
Period: 704.7 minutes

Called STRET 1 by the French, this mission was designed to test various types of solar cells for future French, and perhaps Soviet, spacecraft.

MAS 2

Launch Vehicle : A-2e
Launch Site: Plesetsk
Launch Date: 5 June 1975
Total Weight: 30 kilograms
Apogee: 40,890 kilometers
Perigee: 450 kilometers
Inclination: 63 degrees
Period: 737 minutes

STRET 2 (French designation) was orbited in order to conduct engineering tests of materials designed to provide thermal protection for future payloads. Teflon, kenton and other substances were evaluated.

Meteor

Next to the military photo-reconnaissance program, the Soviet meteorological satellite program is the oldest and most successful Russian space applications effort. The Soviets estimate that their weather satellites save around a billion roubles a year in the areas of agriculture, air and sea travel and fishing. In addition, meteorological satellites have significant military applications for all branches of the Soviet armed services.

While the Soviets did not give their weather satellites the Meteor designator until 1969, a development program under the Kosmos rubric can be traced back to the early 1960s. This section covers both these Kosmos meteorological efforts and the succeeding generations of Meteor spacecraft. It also briefly discusses Soviet efforts to develop a geosynchronous meteorological satellite system.

Kosmos Meteorological Missions

Given the interrelationship between photo-reconnaissance and weather forecasting, it is not surprising that early Soviet recoverable payloads in the Kosmos series appear to have had meteorological as well as military observation missions. Both types of missions are dependent upon remote sensing, and the degree of cloud cover over a given area is a crucial input to an imaging mission—especially for the first generation systems of the Soviet photo-reconnaissance program. Also, the methods and equipment required to orient the imaging vehicle correctly via-a-vis the Earth's surface have equal applicability to both weather and military reconnaissance satellites.

Kosmos 4 (launched 26 April 1962) was the first Soviet photo-reconnaissance satellite and also the first unmanned Soviet spacecraft to return television pictures of clouds. Kosmos 7, 9 and 15 (launched in 1962 and 1963) were all declared in subsequent Soviet announcements to have carried meteorological instrument packages. Kosmos 45, 65 and 92 were also Soviet photoreconnaissance flights that carried infrared and ultraviolet sensors for observing cloud formations.

Two non-military Kosmos flights in 1963, numbers 14 and 23, made direct contributions to the Soviet weather satellite program. Although announced as geophysical missions, these craft actually tested stabilization, orientation and power-supply technology for the first dedicated Kosmos weather satellites.

Between 1964 and 1967 there were five developmental launches of weather satellites from Tyuratam. They all had roughly circular orbits between 625 and 650 kilometers altitude. All were inclined at 65 degrees and had orbital periods of around 97 minutes. The last of the developmental craft was Kosmos 122, which was launched on 25 June 1966 and was the first Kosmos flight that the Soviets declared to have a meteorological mission soon after liftoff.

A switch to Plesetsk for meteorological launches in 1967 signalled the beginning of an operational weather satellite program. Like the developmental launches, these missions used the A-1 booster and had similar circular orbits around 650 kilometers altitude. The inclination, however, was changed to 81.2 degrees, in order to increase the global coverage of the orbit. In all, there were five operational Kosmos weather satellites (sometimes referred to by Western experts as the experimental Meteor phase), ending with the satellite designated Kosmos 226 in mid-1968.

The basic design of the Kosmos meteorological satellites has been carried forward into the various generations of Meteor spacecraft. It consists of a cylindrical body approximately five meters long and 1.5 meters wide. There are two large solar cell panels deployed on opposite sides of the main body. The craft is always oriented with its imaging/sensing package facing toward the Earth's surface. This is called the 'bottom' of the vehicle, and it contains two television cameras; infrared, magnetic and actinometric sensor packages; radio antennas; and control equipment. At the 'top' of the vehicle is a Sun-sensor, and, extending from a boom located roughly in the middle of the main body, there is a steerable ground-plane antenna for returning data to ground stations. The upper half of the spacecraft contains the vehicle's power plant, telemetry equipment and thermal regulation system.

The vehicle operates in a store/dump mode. That is, during its orbital period, the satellite sensors are in constant operation, but the resulting weather data is stored on tape and then down-linked to receiving and processing stations in the Soviet Union when the vehicle is within line-of-sight. Soviet receiving stations are located in Moscow, Novosibirsk and Kharborovsk. The early Kosmos and Meteor satellites were stabilized along three axes by using a series of fly wheels that were driven by electric motors and dampened by on-board electromagnets that interacted with the Earth's magnetic field.

Meteor (1)

The first generation of Meteor-designated satellites comprised some 27 launches. The television (TV) system and the infrared (IR) systems on the Meteor 1 vehicle both scanned a strip some 1000 kilometers wide at a resolution of several kilometers. The TV operated during daylight periods, and the IR during periods of darkness. The actinometric sensor package measured Earth surface temperatures across a 2500-kilometer band.

Meteor 1-5 was the first Soviet weather satellite placed in a roughly 900-kilometer orbit, with period of 102 minutes, allowing better coverage and more accurate geographic location of weather phenomena. This orbit became standard for Meteor spacecraft with Meteor 1-10 in late 1971. Meteor 1-10 was also the first Soviet weather satellite to carry automatic picture transmission (APT) technology that could beam pictures to US as well as Soviet ground stations. Finally, Meteor 1-10 was a testbed for an orbital correction engine that used solar power to create thrust via plasma accelerated through electromagnetic fields. Such ion thrusters are now standard on Meteor satellites. The final Meteor (1) satellite was launched on 5 April 1977.

Meteor (2)

The second generation of Meteor satellites appeared in mid-1975. The vehicle had essentially the same appearance as its predecessor, with the exception of two (rather than one) boom-deployed antennas for data-linking to ground stations. These antennas were mounted near the meteorological package at the 'bottom' of the spacecraft. Major improvements were made, however, in the meteorological instruments. The new package included:

· a telephotometer with APT capability that scanned a 2600-kilometer band at two-kilometer resolution
· a one-kilometer resolution TV system that covered a 2200-kilometer swath

· an infrared sensing package with one-kilometer resolution along a 2600-kilometer swath

Improved Meteor (2)

On 25 March 1982 the Soviets launched the first improved version of the Meteor (2) spacecraft from Plesetsk on an F-2 booster. The improved Meteor (2) vehicles operate in a constellation of three vehicles at a higher altitude (960x940 kilometers, as opposed to 895x855 kilometers) and a different inclination (82.5 degrees vs 81.2 degrees) from their predecessors. The higher orbit and steeper inclination afforded by the F-2 launch vehicle provides better coverage of the earth's equatorial regions. Again, the basic Meteor design does not appear to have changed a great deal, but improvements have been made in the meteorological instrument package. These changes include:

· microwave sensors for measuring moisture content of clouds and the extent of snow and ice cover on the Earth's surface, regardless of weather conditions
· higher-resolution IR scanners capable of direct data downlink to receiving stations

Meteor (3)

Following a 1984 failure, designated Kosmos 1612, the Soviets successfully launched the first third generation Meteor on 24 October 1985. Meteor (3) was described by Tass as being equipped with 'opticalmechanical television equipment, radiometry equipment and instruments for geophysical studies.' Deposited in a 1251×1227-kilometer orbit by an F-2 launched from Plesetsk, Meteor 3-1 is the highest operating Soviet meteorological satellite (it has an inclination of 82.5 degrees and an orbital period of 110 minutes). This orbit successfully closes all the equatorial coverage gaps that remained, even with the improved Meteor (2) orbits.

Future Soviet Geosynchronous Meteorological Satellites

The Soviets have long characterized their space-based meteorological activities as a tripartite effort. The first layer of weather data is provided by manned spacecraft, namely Salyut and Mir, operating between 300 and 400 kilometers altitude. Next came the Meteor craft that now occupy a band between 950 and 1250 kilometers altitude. The final element would be a vehicle at geosynchronous altitude similar to the US/European Geostationary Operational Environmental Satellites (GOES). Despite Soviet agreement to place a meteorological satellite in geostationary orbit under a joint US, European, Japanese and Soviet effort, the USSR has yet to deploy such a vehicle (the United States orbited a GOES spacecraft over the Indian Ocean where the Soviet satellite was to have been placed). Western experts have predicted a Soviet geosynchronous weather satellite every year since the early 1980s, and the consensus is still that one will be launched by the end of the decade. Instruments for the Soviet Geostationary Operational Meteorological Satellite (GOMS) will include a television system that collects data in the visible bands at 1 to 2 kilometers of resolution and in the infrared bands at 5 to 8 kilometers of resolution.

Meteor-Priroda

The Meteor (1) vehicle has also been employed for natural resources remote-sensing missions. Soviet writers have claimed that the Priroda, or Nature, programs of remote sensing save their economy over half a billion roubles annually. When combined with the billion-rouble annual savings from meteorological applications, these results make the Meteor the most cost-effective of all Soviet spacecraft. A typical Tass announcement says the Meteor-Priroda mission is to collect 'operative information on the Earth's natural resources in the interests of the various branches of the USSR's national economy, and to continue testing new forms of information and measuring apparatus and methods of long-distance research on the Earth's surface and atmosphere.'

Like the US LANDSAT program, Meteor-Priroda gathers data that assists the development of the agricultural, natural resources, hydrological and maritime sectors of the economy. In addition, these remote-sensing missions collect data valuable for cartography and geodesy.

While Priroda instruments were probably tested on Meteor 1-18 and 1-25, the first developmental Meteor-Priroda flight occurred in mid-1977. The launch of Meteor 1-28 was unusual in two respects: (1) It marked the return to Tyuratam for the first time in a decade for a Meteor launch and (2) it was the first successful Soviet use of a retrograde Sun-synchronous orbit. This type of orbit allows a satellite to pass over a given point on the planet's surface at the same local time throughout the year. Such an orbit facilitates remote sensing for a number of missions (natural resources, military observation, etc). While employed extensively by US spacecraft for a number of civilian and military applications, the only Soviet Sun-synchronous systems are the Meteor-Priroda craft. One possible explanation for this is the payload-to-orbit penalty that must be paid in order to place a satellite in retrograde orbit (that is, opposing the

Earth's rotation). This penalty may explain the use by the Meteor-Priroda of a 600- to 650- kilometer orbital altitude instead of the 900- and 1200-kilometer orbits employed by Meteor (2) and (3) respectively.

Operational missions probably began in 1980 with the launch of Meteor 1-30, which was referred to by the Soviets as an augmented or improved Meteor-Priroda. This vehicle carried a 600-kilogram package of remote-sensing equipment. Following the replacement of Meteor 1-30 in 1981 by a satellite (Meteor 1-31), which remained in orbit only some 90 days, the Soviets discontinued all retrograde launches for over a year. On 24 July 1983 Kosmos 1484 was placed in Sun-Synchronous orbit, accompanied by a launch

announcement similar to previous Meteor-Priroda flights. Kosmos 1689 was launched to replace Kosmos 1484 in October 1985, marking the first use of the A-1 booster for any Soviet launch in two years (all other former A-1 launched satellites had been switched to the F-2). Why the Soviets changed the designator for their Meteor-Priroda payloads back to the generic Kosmos nomenclature is unclear.

The Meteor-Priroda spacecraft is made up of the basic Meteor vehicle described in the previous section, with a multi-spectral remote-sensing package at the 'bottom' of the craft. The instruments include the following television and scanning collection devices:

· Fragment 2: collects in eight bands (visible and near infrared) along an 85-kilometer swath at 80-meter resolution
· MSU-E: Collects in three bands (visible and near infrared) along a 30-kilometer swath at 30-meter resolution
· MSU-M: Collects in four bands (visible and near infrared) along a 2000-kilometer swath at one-kilometer resolution
· MSU-S: Collects in two bands (visible and near infrared) along a 1400-kilometer swath at 240-meter resolution
· MSU-SK: Collects in four bands (visible and near infrared) along a 600-kilometer swath at 170-meter resolution

The Kosmos/Meteor-Priroda orbit is typically at an altitude of 600 to 650 kilometers, inclined at 98 degrees to the equator with an orbital period of 92 minutes.

Molniya

The workhorse of the Soviet communications satellite network is the Molniya-Orbita system. The peculiarities of Soviet geography, combined with Soviet tardiness in developing geostationary communications satellites, led to the development of a highly elliptical orbit that allowed the establishment of a network of semisynchronous communications satellites. The Molniya, or Lightning, orbit employs a very high apogee (40,000 kilometers) over the northern hemisphere and a very low perigee (400 kilometers) over the southern hemisphere, with inclinations lying between 62.8 and 65.4 degrees and periods of 717.75 minutes. The orbital mechanics associated with such parameters dictate that a satellite will spend the majority of its orbital period near apogee and only a very short time at perigee. Thus for eight hours of a Molniya's roughly twelve-hour orbital period the vehicle will loiter at very useful altitudes for communications over the Soviet Union and North America. In fact, even if the USSR had been able to develop geosynchronous satellites faster, the requirement for a Molniya-type system would remain, because a satellite positioned permanently above the equator cannot serve populations located in the extreme north of the Soviet Union. The fact that the Molniya network was not only maintained but expanded following the successful deployment of Soviet geosynchronous communications satellites beginning in 1974 bears witness to this need.

The first developmental launch in the Molniya program was Kosmos 41 from Tyuratam on an A-2e booster. The first operational launch was made in April of the following year. By the end of 1985 some 110 Molniyas had been successfully launched, and five apparent Molniya failures received the ubiquitous Kosmos label.

Molniyas provide domestic telephone, telegraph and television communications during their period of ascent over the USSR and, by transmitting over the Arctic, during their apogees over North America. While a minimal constellation of three Molniyas could provide continuous 24-hour coverage, the Soviets have historically deployed formations of four and eight vehicles (these are described below). Molniya vehicles transmit to and receive orders from the Orbita network of ground stations, which numbered 100 major and 1000 minor sites in 1983.

The basic Molniya vehicle does not appear to have changed very much over the years. The craft weighs 1600 kilograms and is

Left: A Meteor weather satellite attends the cloudtops in this Soviet mockup. *Below left* and *bottom right* are views of Molniya system communications receiving stations, showing their surrounding flatlands. *Below right:* A Meteor-Priroda georesources satellite guides a Soviet vessel to a bountiful fishing spot.

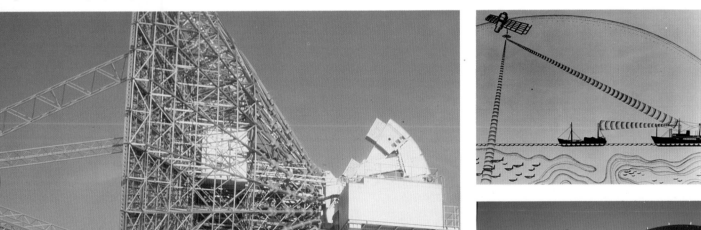

shaped like a flower. The six 'petals' that radiate outward from the body are covered with solar cells that generate 500-700 watts over the typical two-to-three year life of the spacecraft.

The 'stamen and pistil', which constitute the main body of the vehicle, include Sun-seeking optical sensors at one end and Earth-seeking optical sensors at the other. The Sun-seeking apparatus orients the craft so as to optimize the function of the solar cells, and the Earth-seeking sensors maximize the signal strength of the vehicle's transmissions. Between the sensor systems is a pressurized cylinder containing the main instrument package, which includes receivers, transmitters, buffer batteries, a computer and a telemetry system.

Heating and cooling coils are wrapped around the outside of the main body. A special correction motor system and attitude maintenance rockets are also mounted on the main body of the vehicle. Two high-gain, steerable parabolic dish antennas are deployed near the vehicle's base on booms that reach out between two of the solar cell 'petals'. They have a gain of approximately 18 db. Television transmissions are made in the 3400- to 4100-MHz range at a power level of 40 watts. Telephone and telegraph transmissions are made in the 800- to 1000-MHz range at a power level of 120 watts.

The preceding description covers the Molniya 1 system, which is still in operation today. The first Molniya 2 satellite was launched in 1971. Improvements consisted of changes in the power system and the antennas. The solar panels were enlarged, increasing power by 50 percent. Newer, larger antennas were employed, broadcasting at higher (4- to 6-GHz) frequencies. The last Molniya 2 launch occurred in February 1977, and the system is assumed by Western experts to have been phased out. The first Molniya 3 was successfully orbited on 21 November 1974. Its improvements include color television transmission and higher-frequency communications capabilities.

Beginning in the late 1970s, the Soviets maintained a total of 12 Molniyas in orbit, eight Molniya 1s and four Molniya 3s. The Molniya 1s occupied orbital planes spaced 45 degrees apart, thus repeating the same ground tracks every three hours. As discussed elsewhere in this volume, the Molniya 1s are thought to handle Soviet governmental and military communications.

Until 1985 the Soviets maintained four Molniya 3s spaced 90 degrees apart and in the same planes as four of the Molniya 1s. In this configuration the Molniya 3s repeated ground traces every six hours. In 1985 the USSR expanded the Molniya 3 network, which is believed to carry civilian telecom-

munications traffic, to a constellation of eight satellites. The new Molniya 3s have ascending nodes around 155 degrees east. Their purpose is tied to an unspecified 'special program,' according to Soviet launch announcements.

The Molniya orbit affords significant opportunities for geophysical experiments and synoptic observations of the Earth's northern hemisphere. Several Molniya vehicles have carried out scientific measurements in addition to their basic communications missions. Molniya 1-3 and 1-4 were equipped with television cameras that took pictures of cloud cover over the northern hemisphere. In 1971 Molniyas were used to study particle flows in the Earth's Van Allen radiation belts. In October 1975 a Molniya 2 spacecraft collected data on alpha particles and protons.

Oreol

The Oreol series of scientific payloads is a joint Soviet-French program concerning the study of the ionosphere and the aurora borealis. The Oreol flights continued geophysical missions initiated with Kosmos 261 and 348. There have been three Oreol launches to date. These missions are summarized below.

Oreol 1
Launch Vehicle: C-1
Launch Site: Plesetsk
Launch Date: 27 December 1971
Total Weight: 630 kilograms
Apogee: 2500 kilometers
Perigee: 410 kilometers
Inclination: 74 degrees
Period: 114.6 minutes

The payload contained a French magnetometer for measuring low-energy ranges in the auroral region and Soviet instruments for measuring the higher-energy ranges. Ground stations in Bulgaria, Czechoslovakia, East Germany, Hungary, Poland, Romania and the USSR participated in the experiments.

Oreol 2

Launch Vehicle: C-1
Launch Site: Plesetsk
Launch Date: 26 December 1973
Total Weight: 680 kilograms
Apogee: 1995 kilometers
Perigee: 407 kilometers
Inclination: 74 degrees
Period: 109.2 minutes

Left: **This closeup of a Molniya satellite shows *(right to left)* its attitude control sensors, receivers and transmitters and solar panels.** *Above:* **This illustration shows the elliptical orbit and communications scheme of the Molniya 1 satellite—also known as 'Lightning.'**

Basically an identical payload to its predecessor, this mission also explored the possibility of controlled nuclear power plants in orbit.

Oreol 3

Launch Vehicle: F-2
Launch Site: Plesetsk
Launch Date: 21 September 1981
Total Weight: 1000 kilograms
Apogee: 1993 kilometers

Perigee: 399 kilometers
Inclination: 82.5 degrees
Period: 109.5 minutes

Seven years in planning, the third Oreol flight was by far the most ambitious. The 200 kilogram package was contained in an AUOS vehicle (see the Interkosmos 15 entry for details) and was placed in orbit by the modern F-2 booster. Payload weight was divided evenly between French and Soviet instruments. Experiments included measurements of charged particle flows and of magnetic and electric field fluctuations, as well as photometric observation of the northern lights. Oreol 3 marked the first use of foreign equipment for Soviet spacecraft operations. The French provided a telemetry system for direct data transmissions to French receiving stations, a microcomputer for data collation and preliminary processing and an infrared horizon sensor for AUOS orientation.

Polet

The USSR orbited the first maneuverable satellites in the 1960s. Designated Polet 1 and 2, they were launched from Tyuratam with A-m boosters on 1 November 1963 and 12 April 1964, respectively. Polet 1 was hailed by Nikita Khrushchev as representing the dawn of a new era in spaceflight. Both vehicles were evidently maneuvered extensively during their respective 22-day and 57-day missions. One of the results of the program may have been the development of the maneuverable stage employed on the F-class launch vehicle (see Appendix II for details).

Prognoz

Since the early 1970s the Soviets have placed spacecraft bearing the Prognoz label into highly elliptical orbits for the purpose of studying solar radiation (especially the solar wind and solar flares), the Earth's plasma mantle and magnetosphere and terrestrial weather. In addition to scientific data on the Sun, Earth and the origins of the universe, Prognoz flights have contributed to Soviet manned space missions by providing information on solar activity (used to forecast favorable periods for manned missions) and by supplying data on the vulnerability of spacecraft to penetration by solar particles.

All the Prognoz flights to date have employed the same basic vehicle with different instrument packages. The vehicle weighs almost a metric ton and consists of a pressurized cylinder, with four triangular solar panels deployed at right angles to the main body. The satellite is spin-stabilized and has a variety of antennas and other scientific apparatus attached to the hemispherical ends of the main body and to the solar panels. As of 1985 there had been ten Prognoz launches. These missions are described below.

Prognoz 1

Launch Vehicle: A-2e
Launch Site: Tyuratam
Launch Date: 14 April 1972
Total Weight: 845 kilograms
Apogee: 200,000 kilometers
Perigee: 950 kilometers
Inclination: 65 degrees
Period: 4.04 days

The first Prognoz flight studied the interaction of the Earth's magnetosphere with corpuscular, gamma, x-ray and solar plasma radiation. Instruments included five spectrometers (two scintillation, one proton flux, one x-ray and one gamma ray), a Cherenkov electron counter, a solar wind sensor, a magnetometer, orientation detectors, a dosimeter and a radio emission detector (1.6-8 KHz and 100–700 KHz).

Prognoz 2

Launch Vehicle: A-2e
Launch Site: Tyuratam
Launch Date: 29 June 1972
Total Weight: 845 kilograms
Apogee: 200,000 kilometers
Perigee: 550 kilometers
Inclination: 65 degrees
Period: 4.04 days

With essentially the same mission and equipment as its predecessor, Prognoz 2 also carried a French-built apparatus for solar wind experiments.

Prognoz 3

Launch Vehicle: A-2e
Launch Site: Tyuratam
Launch Date: 15 February 1973
Total Weight: 845 kilograms
Apogee: 200,000 kilometers
Perigee: 590 kilometers
Inclination: 65 degrees
Period: 4.01 days

This flight was essentially identical to the first two Prognoz missions.

Prognoz 4

Launch Vehicle: A-2e
Launch Site: Tyuratam
Launch Date: 22 December 1975
Total Weight: 905 kilometers
Apogee: 199,000 kilometers
Perigee: 634 kilometers
Inclination: 65 degrees
Period: 3.98 days

This flight marked the first use of a heavier vehicle, which probably allowed more payload. A redesign of the payload space on the satellite could explain the almost three-year hiatus in the Prognoz program.

Prognoz 5

Launch Vehicle: A-2e
Launch Site: Tyuratam
Launch Date: 25 November 1976
Total Weight: 905 kilograms
Apogee: 199,000 kilometers
Perigee: 510 kilometers
Inclination: 65 degrees
Period: 3.96 days

Prognoz 5 continued experiments concerning the interaction of solar radiation with the Earth's magnetic fields. In addition to the usual suite of Soviet instruments carried into orbit on all the Prognoz craft up to this time, Prognoz 5 had French-built equipment for measuring the solar wind's electron density, as well as the amount of helium and hydrogen present at the altitude reached by the satellite's eccentric orbit.

Prognoz 6

Launch Vehicle: A-2e
Launch Site: Tyuratam
Launch Date: 22 September 1977
Total Weight: 910 kilograms
Apogee: 197,900 kilometers
Perigee: 498 kilometers
Inclination: 65 degrees
Period: 3.95 days

Along with the usual instrument suite, the sixth Prognoz mission included the following equipment:

· Czech instruments for measuring electron density of solar flares
· a French device, called Galactica 1, which collected data on ultraviolet radiation
· Franco-Soviet equipment for measuring the hydrogen and helium concentrations, and the corpuscular content, of solar plasma

Prognoz 6 collected extensive data on solar flares, and this was coordinated with data collected by Prognoz 5 and 7, Signe 3 and Venera 11 and 12 as part of the Solar International Gamma Ray and Neutron Experiments program.

Prognoz 7

Launch Vehicle: A-2e
Launch Site: Tyuratam
Launch Date: 30 October 1978
Total Weight: 910 kilograms
Apogee: 202,965 kilometers
Perigee: 483 kilometers
Inclination: 65 degrees
Period: 4.08 days

Prognoz 7 continued the solar and magnetospheric observation program of its predecessors. New instruments included:

· Bulgarian, Czech and Soviet equipment for studying ion streams in the solar wind
· a Czech Photometer calibrated for x-ray radiation
· the French Galactica 2 device for measuring ultra-violet radiation
· Franco-Soviet equipment for collecting data concerning x-ray, gamma and corpuscular solar radiation
· a Soviet magnetometer and an SKS plasma spectrometer
· the Swedish PROMICS electromagnetic analyzer

Gamma ray experiments conducted by Prognoz 7 were coordinated with similar data collected by Venera 11 and 12, the US Pioneer Venus orbiter and the US/European International Sun-Earth Explorer 3.

Prognoz 8

Launch Vehicle: A-2e
Launch Site: Tyuratam
Launch Date: 25 December 1980
Total Weight: 910 kilograms
Apogee: 199,000 kilometers
Perigee: 550 kilometers
Inclination: 65 degrees
Period: 3.97 days

This mission continued studies of the solar wind and its interaction with the Earth's magnetosphere. The following new scientific equipment was aboard Prognoz 8:

· Czech-Soviet equipment: Two spectroanalyzers, an x-ray solar photometer and a charged particle detector
· a Polish-Soviet spectrometer for ultra-low energy measurements
· a Swedish PROMICS analyzer for charged particles

Prognoz 9

Launch Vehicle: A-2e
Launch Site: Tyuratam
Launch Date: 1 July 1983
Total Weight: 910 kilograms
Apogee: 720,000 kilometers
Perigee: 320 kilometers
Inclination: 65.8 degrees
Period: 26.7 days

This was by far the most ambitious mission of the Prognoz series. Tass announced that the purpose of Prognoz 9 was 'to carry out research into the radio radiation remain-

Right: Preparations for this project may have taken more than 10 years. Prognoz 10 Interkosmos, with an international payload, is seen here immediately before being taken to the pad at the Tyuratam facility.

ing from the moment of the original explosion of the universe, x-ray and gamma flashes in far-out space, as well as corpuscular and electromagnetic radiation of the sun, plasma flows and magnetic fields in near-Earth space to determine the impact of solar activities on the interplanetary medium and the Earth's magnetosphere'. Czechoslovakia, France and the Soviet Union all contributed equipment to the mission, which was designed to use the spacecraft's highly elliptical orbit (the apogee is almost 350,000 kilometers beyond the Moon's orbit) to gather data on the history and composition of the universe.

The main instrument is a radio telescope with two antennas operating at 8mm. One antenna operates along the plane of the elliptic, and one operates perpendicular to the plane of the solar system. Due to the rotation of the satellite and the programmed changes in the vehicle's axis orientation, the radio telescope can map the entire celestial sphere in six months. Soviet sources have reported the telescope's sensitivity as being measured in ten thousandths of a degree of arc.

Prognoz 10

Launch Vehicle: A-2e
Launch Site: Tyuratam
Launch Date: 26 April 1985
Total Weight: 910 kilograms
Apogee: 200,320 kilometers
Perigee: 421 kilometers
Inclination: 65 degrees
Period: 4 days

Planning for this flight reportedly took the better part of a decade. The mission is part of the Intershock Project, which is designed to study the interface between solar plasma and the Earth's magnetosphere. Some 100,000 to 150,000 kilometers from the Earth's surface, on the daylight side, the planet's bow shockwave crosses the energy of solar radiation. This event, measured in seconds, is studied by a Soviet-Czech payload composed of a Czech computer and Soviet plasma sensors.

Progress

In the late 1970s the Soviets modified the Soyuz manned spacecraft design in order to produce a resupply vehicle/space tug for their second generation Salyut space stations. The Progress series of spacecraft is typical of the USSR's approach to its manned space effort in general. The craft represents an incremental, technically

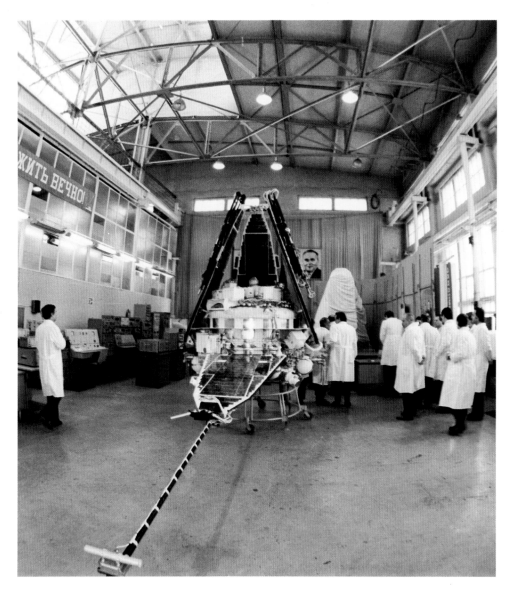

low-risk solution to the problem of maintaining long-duration manned missions in near-Earth orbit. The delivery of supplies, fuel and a capability to adjust Salyut's orbit, however, constitute critical steps in achieving the Soviet goal of a permanent manned presence in orbit.

By the end of 1985 some 24 Progress missions had been flown to Salyut 6 and 7, and, early in 1986, the 26th Progress vehicle docked with the new Mir space station. Without these resupply efforts, the major Soviet space accomplishments of the last few years—such as manned missions exceeding 200 days in orbit and the restoration of Salyut 7 to operational status following a catastrophic control system failure—would have been extremely difficult or impossible. This section describes the Progress vehicle, its operations and the Progress follow-on vehicle.

Soviet engineers essentially stripped a Soyuz vehicle of all equipment needed to support manned space flight in order to maximize the cargo carrying capability of the Progress craft. The removal of all life-support equipment, heat shield, parachute systems and solar panels resulted in a lifting

capability for Progress of 2300 kilograms of supplies and fuel. Fully loaded, the Progress vehicle weighs about 500 kilograms more than a Soyuz configured for manned flight (7020 vs. 6557 kilograms).

The forward end of the Progress vehicle is fitted with a docking collar that has two ports to facilitate refueling of the Salyut space station. Automatic hydraulic connectors are employed to effect the coupling of the two vehicles' main fuel lines. Docking is accomplished using a radar homing system, and redundant warning systems are built into the collar to assure correct interlocking of various latches and connections.

Directly behind the docking collar is the spherical freight module, which contains all non-fuel supplies. The packaged freight is mounted on a special frame and tied down with rapid-release apparatus. Unloading this cargo is somewhat tricky, because it has to be accomplished without upsetting the stability of the three-spacecraft complex (usually a Soyuz craft, a Salyut station and a Progress vehicle). The cargo module usually contains 1300 kilograms of food, water, clothing, experiments, replacement parts and personal items (mail, newspapers, etc).

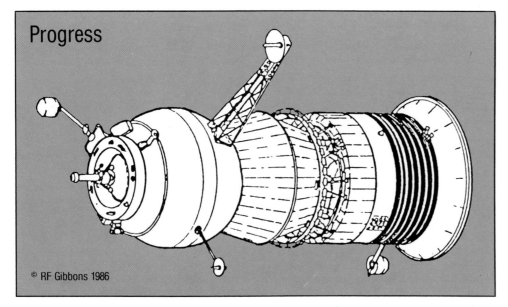

Progress

© RF Gibbons 1986

The refueling complex is located behind the cargo module. It contains four tanks for fuel (hydrazine) and oxidizer. Compressed air and nitrogen for the space station are also stored in this area, along with the equipment necessary to pump the fuel and gases into the Salyut's tanks. Nominal loading for fuel and gases is 1000 kilograms.

Next comes the instrument module, which holds almost twice the amount of control equipment as a basic Soyuz. Behind the instrument compartment is the craft's main engine. External apparatus abound, including:

· three docking lights
· two TV docking cameras
· 14 docking and orientation engines (10 kilograms of thrust each)
· eight precision-orientation engines (one kilogram of thrust each)
· various radio antennas, rendezvous apparatus and sensors

Progress is launched from Tyuratam on an A-2 booster. The docking maneuver is usually carried out a little over 48 hours into the mission. Progress can only dock at the aft Salyut docking port. If a Soyuz craft already occupies this port, it must be moved to the forward docking port in a maneuver involving the undocking of the Soyuz with the crew inside and a rotation of the Salyut around one of its transverse axes, followed by a redocking of the Soyuz at the forward port. The Progress vehicle has sufficient battery power to operate autonomously for eight days and has demonstrated a flight duration capability of two months.

Once a Progress craft has docked with a Salyut station, the crew of the space station unloads the non-fuel supplies and fills the Progress cargo module with refuse. Refueling then takes place in three stages: (1) Empty Salyut tanks are purged of the nitrogen,

which is used to force fuel from the tanks into the space station engines; (2) nitrogen compressed to eight atmospheres is used to pump fuel and then oxidizer, from the Progress fuel module into the Salyut tanks; and (3) all fuel lines are purged, again using nitrogen. Air is also transferred to Salyut to replace that lost each time docking air locks are opened or extra-vehicular activity (EVA) is conducted. Refueling has been conducted with and without a crew present in the space station.

After refueling, Progress is usually employed as a space tug to raise the orbit of the Salyut complex. When the Progress undocks from the space station it is used as a target for testing Salyut's navigation system. The spacecraft is then deorbited, so that it burns up on reentry over the southern Pacific.(For details concerning the Salyut space station see Appendix I. For details concerning the manned Soyuz spacecraft see the Soyuz entry in this volume.)

Progress Follow-On Spacecraft

Four missions of what appear to be Progress follow-on vehicles have been flown since 1977. Kosmos 929 (launched from Tyuratam on 17 July 1977 with a D-1 booster) underwent extensive orbital maneuvers and deorbited a recoverable capsule during its flight. Kosmos 1267 docked with Salyut 6 on 19 June 1981, causing speculation in the Western press concerning a possible anti-satellite/anti-missile capability. (This episode is discussed under the Military Kosmos entry in the subsection on directed energy and advanced kinetic energy ASAT systems.) Kosmos 1443 put these rumors to rest when it docked with Salyut 7 on 10 March 1983, and the Soviets announced that it was an operational follow-on to Kosmos 1267 and that both were members of a new series of super cargo/space tugs. Kosmos 1686 completed the first docking of its class

with an occupied space station when it docked with Salyut 7 on 2 October 1985.

The Progress follow-on appears to be an entirely new type of spacecraft. The entire vehicle weighs 20 metric tons. The main body is shaped something like the US Gemini Program capsules and measures 13 meters in length and four meters in diameter, thus being similar in size to a Salyut station itself. The forward end of the spacecraft is occupied by the descent capsule and its retrorocket system. The reentry system weighs six metric tons and can return 500 kilograms from orbit. Next is the orbital module, weighing some 14 metric tons and capable of carrying three metric tons of cargo (Kosmos 1686, with its descent module removed, delivered five metric tons of cargo and three metric tons of propellant to Salyut 7). Deployed in wing fashion from the main body are two solar cell panels that span 16 meters, cover 40 square meters and provide three kilowatts of power.

The Soviets have announced the following mission roles for this series of spacecraft in the future:

· space station resupply
· cargo return via descent module
· space tug
· space station expansion by 50 cubic meters (for a Salyut-type station this represents a 50 percent increase in living space)
· autonomous operations (eg materials processing)
· modular construction of large, permanently manned orbital complexes (perhaps in conjunction with third generation Mir-type space stations)

Proton

In the mid-1960s the Soviets unveiled their heaviest launch vehicle to date, the D-class or Proton booster. The early launches of this booster carried satellites, called Protons, that were designed to study cosmic rays. While testing the functioning of the booster was probably the main purpose of at least the first three Proton launches, these large satellites did conduct legitimate scientific investigations. The four Proton missions are described in this section.

Proton 1
Launch Vehicle: D-1
Launch Site: Tyuratam
Launch Date: 16 July 1965
Decay Date: 11 October 1965
Total Weight: 12,200 kilograms
Apogee: 627 kilometers
Perigee: 190 kilometers
Inclination: 63.5 degrees
Period: 92.5 minutes

Total Weight: 12,200 kilograms
Apogee: 630 kilometers
Perigee: 190 kilometers
Inclination: 63.5 degrees
Period: 92.5 minutes

This mission measured the energy spectrum and chemical composition of both solar and galactic cosmic rays. Measurements were undertaken in order to attempt detection of quarks and particles with a fractional electron charge.

Proton 4
Launch Vehicle: D-1
Launch Site: Tyuratam
Launch Date: 16 November 1968
Decay Date: 24 July 1969
Total Weight: 17,000 kilograms
Apogee: 495 kilometers
Perigee: 255 kilometers
Inclination: 51.4 degrees
Period: 91.8 minutes

After the third Proton launch the Soviets began using the D-class booster for the Zond series of manned lunar precursor missions (see the entry in this volume for details), but in late 1968 they launched a final Proton cosmic ray spacecraft. An evolution in design, Proton 4 weighed 17 metric tons, with a scientific payload of 12.5 tons. The main body was tapered at one end and conical at at the other. It was nine meters long and had a diameter of 4.5 meters. Deployed, its triangular solar panels spanned 10 meters.

The premier instrument in the huge payload was an ionization calorimeter made of steel bars and plastic scintillators. The measurement device was an instrument composed of one lump of carbon and one of polyethylene. The payload measured the chemical composition and energy ranges of cosmic rays, collected data on the collisions of cosmic ray particles and continued the search for quarks.

Raduga

The Raduga, or Rainbow, geosynchronous communications satellite program became operational in December 1975. The accompanying Tass announcement stated that its function was 'to ensure uninterrupted round-the-clock telephone and telegraph radio communications in the centimetric wave band and simultaneous transmission of color and black-and-white central television programs to the network of Orbita stations'. The Raduga vehicle probably has a cylindrical five-meter-by-two-meter main body, with two solar panels, two antennas, one TV channel and 10 telephonic data channels (each of which car-

The first three Proton spacecraft had a cylindrical main body approximately four meters in diameter. Four panels covered with solar cells folded out from the top of the vehicle. Various antennas also sprouted from the main body. The Soviets announced that the four-metric ton payload, composed of metal, plastic and paraffin blocks, could collect data on cosmic rays with energy levels up to 100 trillion electron volts.

Proton 2
Launch Vehicle: D-1
Launch Site: Tyuratam
Launch Date: 2 November 1965
Decay Date: 6 February 1966
Total Weight: 12,200 kilograms

Above: The large Proton satellite was propelled into space by the then-newly developed D-1 heavy booster. The Proton series investigated quarks and cosmic rays.

Apogee: 637 kilometers
Perigee: 191 kilometers
Inclination: 63.5 degrees
Period: 92.6 minutes

This flight was essentially identical to Proton 1.

Proton 3
Launch Vehicle: D-1
Launch Site: Tyuratam
Launch Date: 6 July 1966
Decay Date: 16 September 1966

Above: In the sunset of an age? The anticipated Soviet space shuttle awaits launching. A streamlined facility forecasts routine usage in this artist's portrayal of a Soviet dream. *See also* pages 40 and 55.

ries 100 multiplexed phone circuits). Soviet sources have mentioned the following features and equipment:

· a three-axis system of fine orientation towards the Earth
· a power supply system with Sun-seeking guidance and tracking of solar cell batteries
· an orbital correction system
· a thermoregulation system
· a radio system for precision measurement of orbital parametric and satellite control

By the end of 1985 some 17 Raduga satellites had been successfully placed in geostationary orbits. The five Raduga 'slots' on the geostationary ring located 35,787 kilometers above the Earth's equator are: 35 degrees E, 45 degrees E, 85 degrees E, 128 degrees E and 335 degrees E. While the primary function of the Raduga network is to supply Soviet domestic telecommunications, a Raduga satellite has supplied service to India since 1983. Raduga platforms will probably be augmented with second generation Soviet communications systems by the

end of the decade. Specifically, the Raduga appears slated to receive the Gals (military), Luch P (general communications) and Volna (air and sea mobile) systems.

Reusable Soviet Spacecraft

Since the mid-1970s there have been persistent rumors that the Soviets are developing reusable spacecraft. The origins for such speculation in the Western press can traced to three basic sources: (1) Soviet statements concerning the utility of reusable manned spacecraft, (2) official statements and leaks to the press from US government sources concerning Soviet efforts to develop reusable vehicles and (3) actual Soviet Kosmos missions designed to develop aspects of technology crucial to the successful deployment of reusable craft. Unfortunately, none of the basic sources has displayed internal consistency over the years. In many respects this confusion may represent an evolving Soviet attitude toward the utility and economic feasibility of reusable spacecraft in a program that has long counted its stable of expendable boosters as a major strength. This section briefly examines the three basic sources for information

concerning the Soviet reusable spacecraft program and their implications for future Soviet initiatives in this area.

Soviet pronouncements concerning plans for reusable spacecraft have been sporadic and contradictory. During a Radio Moscow broadcast in October 1978 a listener posed a question concerning plans for future Soviet spacecraft. In an unusual break with the Soviet tradition of secrecy concerning details of future space efforts, the announcer replied that the USSR was developing a space plane capable of returning from orbit in an aerodynamic mode. Several details concerning the craft's appearance were relayed to the listening audience. The vehicle was said to be shaped like a Delta-winged aircraft and to have three rocket engines at the aft end. The length of the space plane was reportedly 60 meters and the diameter around eight meters.

In 1979 Soviet cosmonauts attending an international conference in Munich indicated that the USSR was exploring the possibility of building a space shuttle system similar to that of the United States, but they qualified these revelations by noting that such spacecraft would have to be developed in such a way as to spare the domestic economy undue burden. A year later another cosmonaut declared that a shuttle system was not presently 'justified' and that expendable boosters could achieve the present Soviet space agenda 'in an economic way.' This statement was followed in 1981 by a complete denial by Cosmonaut Aleksey Yeliseyev that the USSR had any plans whatsoever to build a shuttle. The next year, however, a colleague seemed to reverse this line by indicating that a shuttle was under development but that little would be seen of it for the next two to five years. In August 1983 Konstantin Feoktistov, one of the designers for the Salyut series of Soviet space stations, declared that reusable spacecraft were currently 'unprofitable.' Finally, in an interview with the Western press in 1985, a deputy director of the USSR's Space Research Institute linked the emergence of a Soviet reuseable spacecraft to the deployment of a permanent space station, scheduled for the early 1990s.

US official sources, especially the Department of Defense, have been far more optimistic concerning Soviet reusable spacecraft programs than the Soviets. According to the Pentagon, the USSR is developing both a shuttle and a spaceplane. Rumors have circulated in the Western press since the mid-1970s that the Soviets have been conducting drop tests with a shuttle prototype carried by a M-4 Bison converted medium-range bomber. The tests reportedly take place at the Ramenskoye Aircraft Experimentation Center. The Soviet shuttle is said to be similar to the US orbiter but differs from the American system in two key aspects. In place of rocket engines on the vehicle itself, the Soviets have installed jet engines. This scheme allows a greater degree in flexibility with respect to landing the vehicle following reentry. The Soviet shuttle cannot take off under the power of its jet engines, but it could undertake emergency maneuvers to correct its approach path to a runway, should it get off course during reentry, or it could go around for a second approach if the first were waved off by ground control. Neither of these options is available with the US orbiter. The penalty paid by foregoing rocket engines on the vehicle itself is that all the launch boosters must be expendable. The other major difference between the two orbiters is that the Soviet system is all liquid fuel, while the US orbiter is launched with a combination of liquid and solid fuel boosters.

The following dimensions have been reported in the West: a 23-meter wing span, 33-meter length and a 5.5-meter diameter. The Pentagon estimates that the Soviet shuttle can place a payload of 30 metric tons into low-Earth orbit. A long runway built recently at Tyuratam has also been linked to the Soviet shuttle program. The Defense Department estimates that the first Soviet shuttle

launch will be conducted in late 1986 or early 1987, but other Western experts believe that the Soviets are much further away from such a momentous step.

The USSR is reported to be close to operating its spaceplane system as well. This speculation is tied to the series of tests involving the reentry of subscale spaceplane models. In general, the Soviet spaceplane is thought to resemble the US Air Force's X-20 (Dyna-Soar) vehicle, which was cancelled in the early 1960s. Dyna-Soar was envisioned as a delta-winged reusable vehicle with a multi-member crew and was to have been used as an orbital observation or weapons platform.

Unfortunately, Soviet Kosmos launches that might be related to reusable spacecraft development shed very little light on the continuing mystery. Such flights fall into two categories: the Kosmos pairs of the mid-to-late 1970s and the subscale spaceplane model tests of the early-to-mid 1980s. The Kosmos pairs were launched on D-1 boosters from Tyuratam. In all, three pairs of vehicles were launched between late 1976 and the spring of 1979 (Kosmos 881/882 on 16 December 1976; Kosmos 997/998 on 30 March 1978; and Kosmos 1100/1101 on 22 May 1979). The mission profiles for all the flights were nearly identical: the vehicles were inclined at 51.6 degrees to the equator, apogees were between 250 and 230 kilometers, perigees were around 200 kilometers and each craft was deorbited before the completion of a single revolution. The press quoted US government sources as describing the first pair as 'definitely man-related.' Other Western specialists linked the flight profiles to tests of a maneuverable vehicle.

The Soviet subscale vehicle tests began on 3 June 1982 with the launch of Kosmos 1374. The 900-kilogram vehicle was launched from Kapustin Yar on a C-1 booster into a 204x158-kilometer orbit, inclined at 50.67 degrees, with an orbital period of 88.1 minutes. The craft was recovered, after only 1.25 revolutions, by Soviet naval vessels in the Indian Ocean. A second, quite similar test was conducted on 15 March 1983. This time the presence of a Soviet naval flotilla composed of seven ships attracted the attention of the Australian Air Force, which managed to film the recovery.

The subscale vehicle indeed appeared to be a miniature of a potentially reuseable spacecraft. The triangular-shaped craft was fitted with a wraparound cockpit and wings angled upward from the main body at 22.5 degrees. A dorsal stabilizer ran along a portion of the vehicle's spine at the rear end. It measured 3.4 meters in length, 1.4 meters in width and had a wing span of some 2.6 meters. The top of the vehicle was partially

covered with what appeared to be ceramic tiles, like those on the US shuttle, for heat protection during reentry, but the bottom of the craft apparently was smooth—perhaps indicative of a single piece of ceramic or some other form of protection. Western experts have estimated that a full-scale vehicle of this type might weigh some 18,000 kilograms (for comparison purposes, the US shuttle weighs 106,000 kilograms).

Perhaps in order to preclude further Western observation of their recovery activities, the Soviets deorbited the next two flights so that they landed in the Black Sea. The mission profiles of both Kosmos 1517 (launched 27 December 1983) and Kosmos 1614 (launched 19 December 1984) were nevertheless identical to those of their predecessors. Confusion exists in the West as to whether the Soviets are developing a specific type of vehicle or are simply exploring a certain area of technology—or whether they are flying several vehicles or the same one over again. Potential missions for the spaceplane may include its use as a space ferry for taking crews to and from space stations, a military observation vehicle or a platform for anti-satellite or anti-missile systems.

No subscale model tests took place in 1985. It is quite possible that the Soviets may still not have decided whether to make the large commitment in resources required to deploy an operational fleet of reusable spacecraft. Soviet space planners clearly face a dilemma, in that while it is tempting to continue to exploit the USSR's proven capability in expendable boosters, the overall thrust of the Soviet program towards a large manned presence in space and the exploitation of near-Earth space for economic benefit will eventually require reusable spacecraft in order to remain viable.

Sneg

Sneg 3
Launch Vehicle: C-1
Launch Date: 17 June 1977
Decay Date: 20 June 1979
Total Weight: 102 kilograms
Apogee: 519 kilometers
Perigee: 459 kilometers
Inclination: 50.7 degrees
Period: 94.3 minutes

The French Signe satellite, called Sneg 3 by the Soviets, carried 28 kilograms of instruments designed to measure the ultraviolet content of the solar wind and to locate discrete sources of x-ray and gamma ray radiation. Data from the payload was downlinked to the Soviet Union and then transferred to France in digital format over telephone lines.

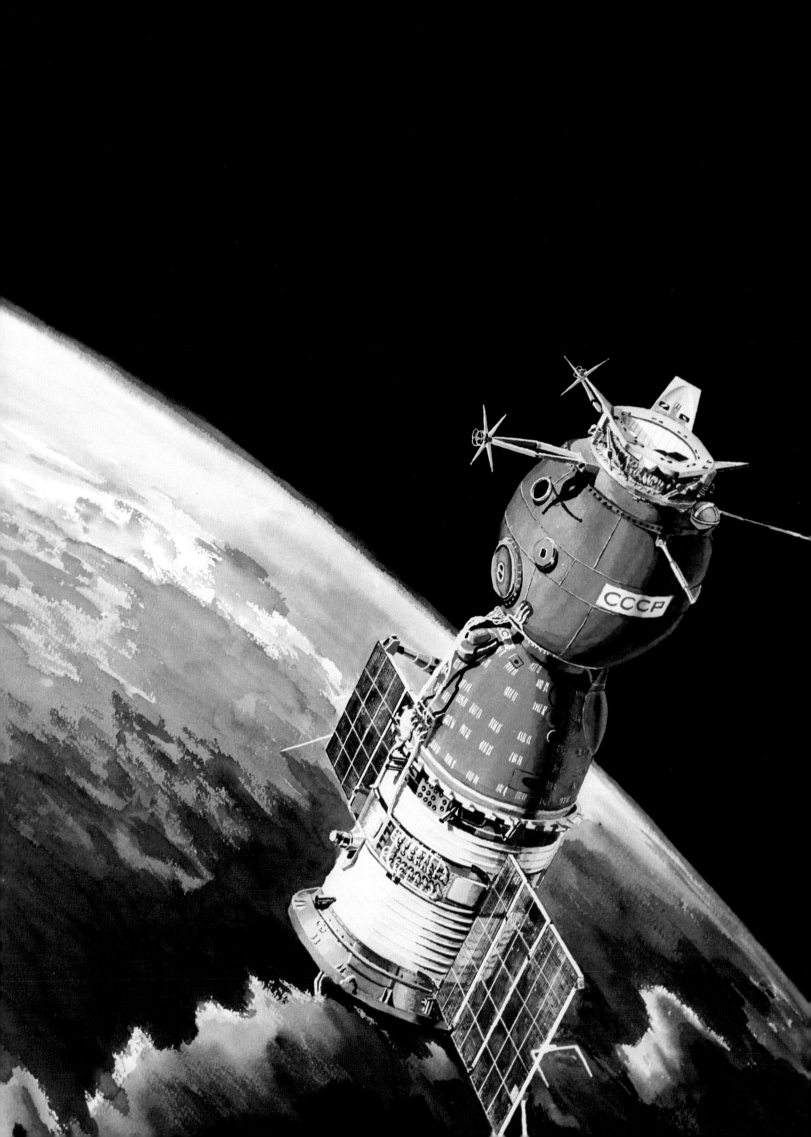

Soyuz

Despite a rocky start, intermittent technical difficulties and the loss of four cosmonauts over the course of the past 20 years, the Soviet manned Soyuz, or Union, program has evolved into a versatile and reliable means of placing humans in near-Earth orbit for a variety of scientific—and probably military—missions. Soyuz constitutes one of the best examples of the Soviet incremental approach to space exploration. By the end of 1985 the USSR had successfully orbited 10 variants of the basic Soyuz vehicle. In early 1986, a Soyuz-T variant carried the first crew to the new Mir space station, and there is no sign of Soviet intentions to abandon the craft which has been the workhorse of their manned space program for the last two decades. This section describes the basic Soyuz design, chronicles the various Soyuz missions (including the Kosmos precursors) and notes spacecraft modifications as they occurred during the evolution of the program.

All Soyuz launches are conducted with the A-2 booster from Tyuratam. The basic vehicle is composed of three segments: the orbital, descent and instrument modules. The egg-shaped orbital module weighs approximately one metric ton, houses life support equipment and consumables and provides some crew space in addition to the descent module. The forward end of the orbital module can be configured for a variety of different missions (eg docking, remote sensing, etc), and the aft end is connected to the descent/command module. The bell-shaped descent module can be configured for a crew of one, two or three members. The reentry capsule was designed with aerodynamic capabilities for precision landings at pre-selected sites in the USSR. Deorbit and recovery are usually conducted at a force of three to four G's, but emergency ballistic reentries up to 10 G's are possible. The final component, the instrument module, provides maneuver, attitude and thermal control. Electric power on all but two Soyuz variants is generated by two sets of rectangular solar panels attached to, and deployed at right angles from, the instrument module. This component also contains the retrorocket engines (two 400-kilogram-thrust liquid fuel systems) that deorbit the craft or change its orbital parameters.

A basic mission profile consists of lift-off from Tyuratam, followed by jettison of the four strap-on stages of the A-2 (see Appendix II for details) at the 120-second mark. The protective shroud covering the Soyuz vehicle is jettisoned 160 seconds after launch. Final-stage ignition occurs 300 seconds into the mission, and orbital insertion at the 530-second mark. Reentry is initiated by a retrorocket burn, ten minutes after which the orbital and instrument models are jettisoned. An 18- to 25-minute blackout follows as reentry occurs. At a height of nine kilometers a drogue parachute deploys, followed shortly by the main chute. When the spacecraft descent slows to six meters per second the heat shield is jettisoned, revealing a retrorocket system in the base of the capsule. This system fires just before impact, at a height of approximately one meter, in order to soften the landing further.

Kosmos 133
Launch Date: 28 November 1966
Recovery Date: 30 November 1966
Total Weight: 6450 kilograms

A 22-month gap occurred between Voskhod 2 and further Soviet man-related space launches. Kosmos 133 was identified as a manned precursor by its radio beacon frequency, low perigee, almost circular orbit and the fact that it was recovered after only two days of flight rather than after the eight day mission which was the standard military photo-reconnaissance profile at the time.

Kosmos 140
Launch Date: 7 February 1967
Recovery Date: 9 February 1967
Total Weight: 6450 kilograms

This flight was almost identical to Kosmos 133, and it constituted the final checkout mission before Soyuz 1.

Left: The sequence of a Soyuz mission. *Far left:* an artist's conception of Soyuz with passive docking apparatus at upper end of craft. *Below:* The Soyuz modules.

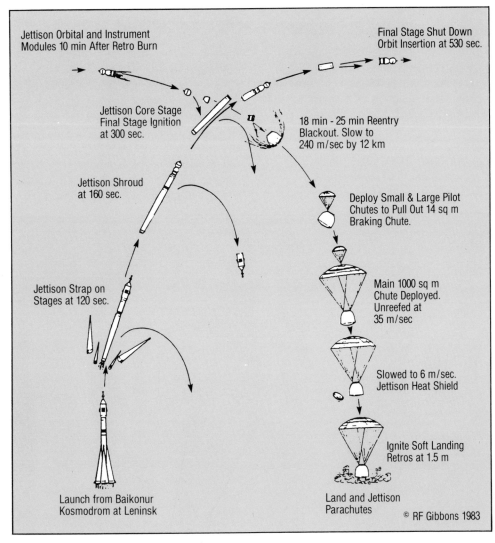

Jettison Orbital and Instrument Modules 10 min After Retro Burn

Final Stage Shut Down Orbit Insertion at 530 sec.

Jettison Core Stage Final Stage Ignition at 300 sec.

18 min - 25 min Reentry Blackout. Slow to 240 m/sec by 12 km

Jettison Shroud at 160 sec.

Deploy Small & Large Pilot Chutes to Pull Out 14 sq m Braking Chute.

Jettison Strap on Stages at 120 sec.

Main 1000 sq m Chute Deployed. Unreefed at 35 m/sec

Slowed to 6 m/sec. Jettison Heat Shield

Ignite Soft Landing Retros at 1.5 m

Launch from Baikonur Kosmodrom at Leninsk

Land and Jettison Parachutes

© RF Gibbons 1983

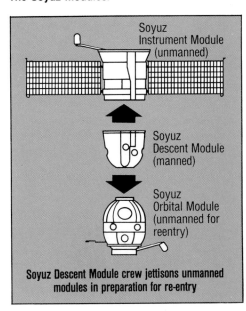

Soyuz Instrument Module (unmanned)

Soyuz Descent Module (manned)

Soyuz Orbital Module (unmanned for reentry)

Soyuz Descent Module crew jettisons unmanned modules in preparation for re-entry

Soyuz 1

Launch Date: 23 April 1967
Landing Date: 24 April 1967
Total Weight: 6570 kilograms
Crew: V Komarov

Despite rumors in Moscow that another Soviet space spectacular was in the offing and a routine Tass announcement that the first Soyuz mission was preceding according to plan, it became clear that something was amiss when this flight was terminated after only 24 hours. Reentry appeared to be normal, but the parachute system failed to deploy properly. Colonel Komarov was killed, and his spacecraft was destroyed on impact. Komarov was the first Soviet cosmonaut casualty resulting from a mission failure.

Kosmos 186

Launch Date: 27 October 1967
Recovery Date: 31 October 1967
Total Weight: 6530 kilograms

The Soviets did not send another manned flight into space for 18 months following the Soyuz 1 tragedy, but unmanned developmental flights continued. Kosmos 186 was the active vehicle in an unmanned automated docking exercise with Kosmos 188. The two craft rendezvoused and docked on the other side of the world from the USSR. Then they completed two and a half orbits before an automatic decoupling sequence was initiated over the Soviet Union.

Kosmos 188

Launch Date: 30 October 1967
Recovery Date: 2 November 1967
Total Weight: 6530 kilograms

Kosmos 188 made a direct-ascent first-orbit rendezvous with Kosmos 186 that brought the two spacecraft within 24 kilometers of each other before Kosmos 188 initiated the docking sequence.

Kosmos 212

Launch Date: 14 April 1968
Recovery Date: 19 April 1968
Total Weight: 6530 kilograms

Another initial vehicle of a docking pair, Kosmos 212 again played the active role. This time, according to the Soviets, the docking took place over Soviet territory, and joint flight lasted almost four hours. Following decoupling, Kosmos 212 and 213 practiced group flight maneuvers.

Kosmos 213

Launch Date: 15 August 1968
Recovery Date: 20 April 1968
Total Weight: 6530 kilograms

Kosmos 213 made a first-orbit direct-ascent link-up with Kosmos 212. The two spacecraft came within five kilometers of each other, with a difference in velocity of 108 kilometers per hour, before the initiation of the docking sequence by the Kosmos 212 spacecraft.

Kosmos 238

Launch date: 28 August 1968
Recovery Date: 1 September 1968
Total Weight: 6520 kilograms

This mission was the final checkout flight before the resumption of manned missions with Soyuz 2 and 3.

Soyuz 2

Launch Date: 25 October 1968
Recovery Date: 28 October 1968
Total Weight: 6520 kilograms

One of only four unmanned flights to carry the Soyuz designator, this spacecraft served as a docking target for Soyuz 3.

Soyuz 3

Launch Date: 26 October 1968
Recovery Date: 30 October 1968
Total Weight: 6575 kilograms
Crew: G Beregovoi

Soviet mission controllers brought Soyuz 3 to within 200 meters of Soyuz 2, after which Colonel Beregovoi assumed manual control for the docking sequence. Although he was able to approach to within one meter of his target, the pilot could not manage to dock with the other spacecraft. These technical difficulties presaged intermittent docking problems that were to plague several future Soyuz missions. The rest of the flight was used to check out the Soyuz flight instrumentation, take photographs and gather geophysical data.

Soyuz 4

Launch Date: 14 January 1969
Recovery Date: 17 January 1969
Total Weight: 6625 kilograms
Crew: V Shatalov

Soyuz 4 was the first manned launch conducted by the Soviets in winter. Previously, Soviet mission specialists had feared that an aborted or unscheduled landing on the steppes in winter would cause delays in reaching the crew, who could be endangered by exposure to severe weather. Both Soyuz 4 and Soyuz 5 were equipped for an ocean landing in warmer climates as a contingency against such a scenario. Soyuz 4 was the active partner in the first successful Soviet manned docking sequence, which was carried out in conjunction with Soyuz 5.

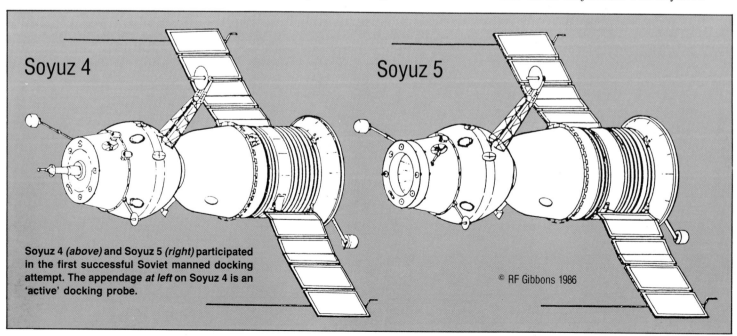

Soyuz 4 *(above)* **and Soyuz 5** *(right)* **participated in the first successful Soviet manned docking attempt. The appendage** *at left* **on Soyuz 4 is an 'active' docking probe.**

© RF Gibbons 1986

Soyuz 5

Launch date: 15 January 1969
Recovery Date: 18 January 1969
Total Weight: 6585 kilograms
Crew: Ye Khrunov, B Volynov and A Yeliseyev

Automatic systems brought Soyuz 5 to within 100 meters of Soyuz 4, after which Colonel Shatalov completed the docking maneuver in manual regime. Once the craft were secured, Cosmonauts Khrunov and Yeliseyev put on space suits, entered Soyuz 5's orbital module and exited the craft. Their EVA terminated in their transfer to Soyuz 4 via handrails installed on the sides of both vehicles and in their use of Soyuz 4's orbital module as an airlock. The crew switch took about an hour and was recorded by television cameras. The Soviets immediately proclaimed their newest accomplishment to be the first establishment of an orbital space station. If so, it was extremely short-lived, as the two craft separated after four and a half hours of joint flight. Soyuz 4 returned with its newly expanded crew one day ahead of Soyuz 5.

Soyuz 6

Launch Date: 11 October 1969
Recovery Date: 16 October 1969
Total Weight: 6577 kilograms
Crew: G Shonin and V Kubasov

The first of three successive daily launches, Soyuz 6 carried out experiments associated with vacuum welding. Such investigations were geared toward judging the feasibility of eventually erecting large, permanently manned orbital complexes. By remote control the crew operated a welding station which was set up in the orbital module. The module was opened to space, and the welding device (called Vulcan by the Soviets) was used to test compressed arc, electrode arc and electron beam welding methods. The electron beam method appears to have produced the best results. The Soyuz 6 crew also conducted further tests of the Soyuz flight systems and gathered data to aid in the exploitation of natural resources.

Soyuz 7

Launch Date: 12 October 1969
Recovery Date: 18 October 1969
Total Weight: 6646 kilograms
Crew: A Filipchenko, V Gorbatko and V Volkov

Soyuz 7 was the intended passive docking partner for Soyuz 8. The crew also gathered data concerning Earth resources.

Soyuz 8

Launch Date: 13 October 1969
Recovery Date: 18 October 1969
Total Weight: 6646 kilograms
Crew: V Shatalov and A Yeliseyev

With the placement of Soyuz 8 in orbit, the Soviets could claim another first, with three spacecraft and seven men in orbit simultaneously. The achievement must have fallen somewhat short of the hopes of the Soviet space planners and mission controllers, however, since, despite extensive maneuvers, Soyuz 8 could not manage to dock with Soyuz 7. While the Soviets claimed at the time that it was never their intention for the two craft to dock, orbiting two vehicles equipped with docking collars and flown by crews experienced in such maneuvers seems more than a little suggestive of intentions to carry out an orbital docking sequence.

Right: **Soyuz 9 atop its launch vehicle on the pad at Tyuratam.**

Soyuz 9

Launch Date: 1 June 1970
Recovery Date: 19 June 1970
Total Weight: 6500 kilograms
Crew: A Nikolayev and V Sevastyanov

This 18-day mission broke the previous record of the US Gemini 7 mission of 14 days, which had stood for five years. The major mission objective of Soyuz 9 was to gather data on the effects of long-term space flight on humans and other organisms. Each

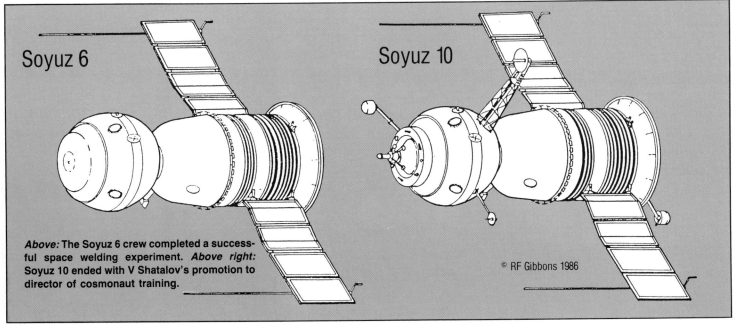

Above: The Soyuz 6 crew completed a successful space welding experiment. *Above right:* Soyuz 10 ended with V Shatalov's promotion to director of cosmonaut training.

© RF Gibbons 1986

crewman monitored the other's bodily reactions to prolonged weightlessness and took computerized tests designed to measure his physical and mental adjustment to a zero gravity environment. Other biological experiments involved the study of the ability of insects, plants and bacteria to adapt and propagate in orbit.

The long mission also afforded time for earth resources observation, navigation exercises and astrophysical experiments. The crew took both black-and-white and multispectral photographs of the Earth in order to locate mineral reserves and schools of fish, determine the moisture content of glaciers and estimate timber reserves. The cosmonauts used a sextant for astronavigation by determining the relation of stars to the Earth's horizon, thereby deriving precise location and orbital data for their spacecraft. Astrophysical observations included taking photographs of the moon. Although Nikolayev and Sevastyanov appear to have been slower to recover from their flight than the Gemini 7 astronauts, they apparently suffered no long-term side effects. This mission probably bolstered Soviet confidence in long duration space flights and paved the way for the marathon Salyut missions of the late 1970s and early 1980s.

Soyuz 10
Launch Date: 22 April 1971
Recovery Date: 24 April 1971
Total Weight: 6575 kilograms
Crew: N Rukavishnikov, V Shatalov and A Yeliseyev

Soyuz 10 was launched for the purpose of docking with the Salyut 1 space station, which had been placed in orbit on 19 April 1971. The crew was mostly made up the veterans of Soyuz 4/5 and 8. At the time, Colonel Shatalov clearly had more docking experience than any other Soyuz pilot, while Flight Engineer Yeliseyev had participated in the EVA between Soyuz 4 and 5. The rookie on the flight, Rukavishnikov, was an expert on the Salyut station's systems. Mission control brought the spacecraft to within 180 meters of the space station, after which Shatalov completed the docking sequence manually.

At this point in the mission things apparently started to go wrong, because Soyuz undocked from the Salyut station after only five and a half hours and made the earliest reentry possible that would allow a landing on Soviet territory. As usual, the Soviets claimed that the mission had gone according to plan, but the failure of the

Cosmonauts exit their 'office', the Soyuz orbital module (left), and enter the Soyuz descent module (not shown) before reentry.

crew—which had a Salyut specialist and a ship-to-ship transfer veteran on board—to enter Salyut 1 is an indication that something was seriously amiss. The additional fact of the obviously hasty reentry (Soyuz 10 was the first Soviet predawn landing) pointed to the Soyuz vehicle as the source of the problem. Given endemic Soviet secrecy, the exact nature of this flight's difficulties will probably never be known, but in hindsight, Soyuz 10 presaged danger for the next mission.

Soyuz 11
Launch Date: 6 June 1971
Recovery Date: 29 June 1971
Total Weight: 6790 kilograms
Crew: G Dobrovolsky, V Patsayev and V Volkov

The second Soviet space tragedy occurred at the end of what must have seemed a very successful mission. Soyuz 11 docked with Salyut 1 on 7 June, and the crew set a new duration record of 23 days while conducting biological, astrophysical and geophysical experiments. Following the undocking sequence and a nominal retrorocket firing for reentry, however, communications with the descent capsule ceased prior to the usual blackout when a spacecraft enters the ionosphere. Apparently a valve malfunctioned at the moment of separation from the orbital module, and all the air escaped from the capsule. The design of the Soyuz reentry module at the time did not allow enough room for three cosmonauts to wear spacesuits. The crew's efforts to close the valve manually did not succeed in time. The three cosmonauts were found dead in their capsule by the Soviet recovery team following touchdown.

Kosmos 496
Launch Date: 26 June 1972
Recovery Date: 2 July 1972
Total Weight: 6570 kilograms

A year after the Soyuz 11 disaster the Soviets began testing an improved version of the spacecraft. This mission was covered by the ubiquitous Kosmos label, but the vehicle's orbital parameters and telemetry formats led Western experts to classify the mission as an unmanned Soyuz test flight.

Kosmos 573
Launch Date: 15 June 1973
Recovery Date: 17 June 1973
Total Weight: 6570 kilograms

This was the final unmanned checkout flight before the resumption of manned missions with Soyuz 12. Kosmos 573's orbital

elements and telemetry resembled those of Kosmos 496, but its mission length of two days foretold the flight duration planned for Soyuz 12.

Soyuz 12
Launch Date: 27 September 1973
Recovery Date: 29 September 1973
Total Weight: 6570 kilograms
Crew: V Lazarev and O Makarov

Soyuz 12 unveiled the precautions that the Soviets developed in the wake of the Soyuz 11 tragedy. The valve systems were almost certainly reengineered and equipped with redundant backups. More important was the introduction of spacesuits. For the next seven years, until the advent of the Soyuz-T variant, the Soviets flew only two-man missions, so that both cosmonauts could wear spacesuits during lift-off and reentry, and to this day, all Soviet cosmonauts wear spacesuits in their Soyuz craft. Soyuz 12 was also configured with a new launch escape rocket. The vehicle itself was a new Soyuz variant designed to ferry cosmonauts from Earth to a Salyut station. The major change involved removal of the solar panels in order to achieve weight savings. Use of internal batteries limited free flight missions to two days.

The primary mission objective was to shake down the new vehicle in preparation for the resumption of manned Salyut missions. The crew also obtained multi-spectral photography of the Earth's surface using a nine-objective camera. These pictures were then compared with those taken of the same area simultaneously by aircraft, in order to study means of reducing atmospheric distortion.

Kosmos 613
Launch Date: 30 November 1973
Recovery Date: 29 January 1974
Total Weight: 6570 kilograms

Kosmos 613 apparently sought to test Soyuz' capability to withstand long-term flight in a powered-down configuration and then to undergo reactivation and recovery. Such capabilities were required for the extended Salyut missions that began with Soyuz 18.

Soyuz 13
Launch Date: 18 December 1973
Recovery Date: 26 December 1973
Total Weight: 6680 kilograms
Crew: P Klimuk and V Lebedev

In the wake of the failure of Salyut 2 and Kosmos 557 (see Appendix I for details) the Soviets returned to an anonymous Soyuz

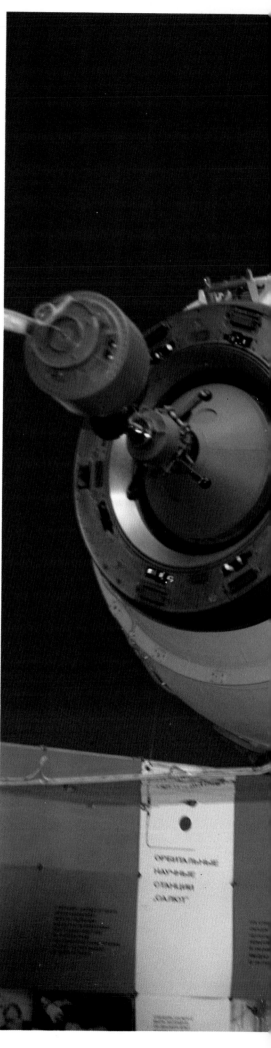

Above: Soviet and American Apollo-Soyuz Test Project personnel trained together at the Yuri Gagarin Training Center. *Below:* Soyuz 3 is here enroute to the launching pad.

A Soyuz spacecraft, showing, *left to right:* the orbital module, the descent module, and the instrument module. Soyuz design is similar to the two-module Vostok craft.

mission with a solar powered craft. Soyuz 13 conducted a multi-faceted eight-day scientific mission. Its primary objective was astronomy, using the Orion-2 ultra-violet camera complex. This three-axis-stabilized system was mounted at the forward end of the orbital module, which itself was converted from a sleeping and leisure area into a laboratory. Orion-2 was used to observe stars that are normally obscured by the Earth's atmosphere and to study solar x-rays. Some 10,000 spectrograms of 3000 stars were made during the mission. In addition to astronomy, the crew of Soyuz 13 conducted biological experiments concerning both synthetic production of biomass and the adaptation of higher order plants to weightlessness. Multi-spectral photography of the Earth's surface and spectrographic atmospheric measurements were also conducted. Finally, the crew engaged in autonomous navigation exercises, building on the experience of Soyuz 9.

Kosmos 638
Launch Date: 3 April 1974
Recovery Date: 13 April 1974
Total Weight: 6800 kilograms

Kosmos 638 was the first of two unmanned precursor flights for the Apollo-Soyuz Test Program (ASTP). The vehicle was placed in the same orbit planned for the joint US-Soviet venture, and it used teleme-

Left: **An artist's conception of the Apollo-Soyuz 19 docking rendezvous.** *Below:* **Soyuz 19.**

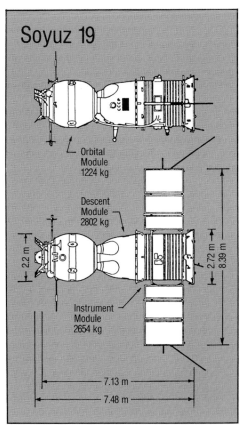

try formats usually associated with manned missions.

Kosmos 656
Launch Date: 27 May 1974
Recovery date: 29 May 1974
Total Weight: 6570 kilograms

Kosmos 656 was another test of the Soyuz ferry variant. The mission was similar to that of Kosmos 573.

Soyuz 14
Launch Date: 3 July 1974
Recovery Date: 19 July 1974

Total Weight: 6570 kilograms
Crew: Y Artyukhin and P Popovich

The first operational mission using the Soyuz in ferry mode, Soyuz 14 remained docked with Salyut 3 for 16 days. The Soviets announced that Soyuz 14 had been equipped with a water recovery capability.

Kosmos 670
Launch Date: 6 August 1974
Recovery Date: 9 August 1974
Total Weight: 6800 kilograms

Kosmos 670 caused a stir in the West when it was placed in an orbit with an inclination of 50.6 degrees, an inclination that had never been used by a vehicle launched by an A-class booster. At the time, some experts thought this mission presaged the imminent launching of the long-expected Soviet heavy G-class booster. In reality, the mission was probably the first test of the Soyuz-T variant.

Kosmos 672
Launch Date: 12 August 1974
Recovery Date: 18 August 1974
Total Weight: 6800 kilograms

Kosmos 672 was the second unmanned ASTP precursor.

Soyuz 15
Launch Date: 26 August 1974
Recovery Date: 28 August 1974
Total Weight: 6570 kilograms
Crew: L Demin and G Sarafanov

Left: Pyotr Klimuk and Valentin Lebedev *(at right)* **of Soyuz 13. Soyuz was redesigned following Soyuz 11 to provide the added protection of spacesuits—originally eliminated to provide room for three men.**

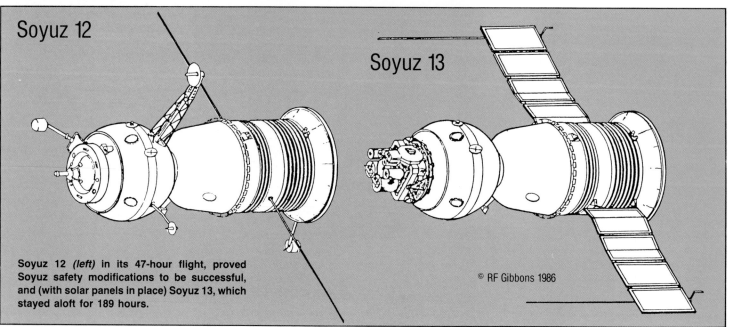

Soyuz 12 *(left)* **in its 47-hour flight, proved Soyuz safety modifications to be successful, and (with solar panels in place) Soyuz 13, which stayed aloft for 189 hours.**

© RF Gibbons 1986

Intended to continue the work of Soyuz 14 inside the Salyut 3 space station, this mission had to be terminated when the vehicle's malfunctioning automatic docking program repeatedly introduced too much thrust into the final closing sequence. Because of the Soyuz ferry variant's limited mission life, the vehicle was forced to make a night landing in bad weather near Tselinograd. Subsequent Soviet claims that the mission was never really intended to make a long stay at the space station, or even to enter the complex, have been dismissed by most Western experts.

Soyuz 16
Launch Date: 2 December 1974
Recovery Date: 8 December 1974
Total Weight: 6680
Crew: A Filipchenko and N Rukavishnikov

Soyuz 16 was a manned precursor ASTP flight using the prime back-up crew for the mission eventually flown by Soyuz 19. The primary mission objective was to test the various modifications and special equipment required to link-up the two technically dissimilar spacecraft. After achieving the 225-kilometer circular orbit planned for the ASTP rendezvous, Soyuz 16 carried out some 20 tests associated with the planned docking sequence by deploying a special ring that simulated the effect of docking with an Apollo vehicle. US tracking stations around the world participated in this mission in preparation for the ASTP flight. The Soyuz 16 crew also assessed the impact of lowering the normal vehicle cabin pressure and increasing the percentage of oxygen in their breathing mixture—measures required in order to facilitate adjustment with Apollo's atmospheric pressure and air composition.

The cosmonauts also had time to conduct a series of biological, astrophysical and Earth-resources experiments. Biological studies included growing microorganisms, fungi, plants and fish in a gravity-free environment. Photographs were taken of the Sun and other stars. Finally, the crew photographed the Earth, continuing previous Soyuz missions' efforts to collect data concerning exploitation of Soviet natural resources and to further the study of the composition and density of the Earth's atmosphere.

Soyuz 17
Launch Date: 10 January 1975
Recovery Date: 9 February 1975
Total Weight: 6570 kilograms
Crew: G Grechko and A Gubarev

Soyuz was the first ferry mission to Salyut 4. Landing and recovery were complicated by a serious snow storm, which produced very limited visibility (500 meters), and by high winds (20 meters per second). Nevertheless, a Soviet helicopter was able to remove the cosmonauts from the landing site within ten minutes of touchdown.

Soyuz (unnumbered)
Launch Date: 5 April 1975
Recovery Date: 5 April 1975
Total Weight: 6570 kilograms
Crew: V Lazarev and O Makarov

This mission was forced into a sub-orbital abort when the A-2 launch vehicle underwent a stage separation malfunction. The crew landed in Siberia some 320 kilometers from the Sino-Soviet border. The extreme cold forced the cosmonauts to leave their vehicle and build a fire. The crew were soon recovered in good health, but the occurrence

of a major Soviet launch failure only three months before the scheduled ASTP mission caused some anxiety in the US. While some American scientists and legislators called for a postponement or cancellation of the ASTP mission, citing what the Soviets were by that point terming 'the April anomaly,' NASA took the decision to continue with the programmed launch.

Soyuz 19
Launch Date: 15 July 1975
Recovery Date: 21 July 1975
Total Weight: 6800 kilograms
Crew: V Kubasov and A Leonov

On 17 July 1975 Soyuz 19 rendezvoused with an Apollo spacecraft, docked and conducted two days of joint scientific experiments. The Apollo vehicle was the active partner in the docking sequence, and the docking mechanism was of joint US-Soviet design and US manufacture. The Apollo craft carried a crew of three: Thomas Stafford, Vance Brand and Donald Slayton. General Stafford and Colonel Leonov entered the docking mechanism airlock and shook hands over Metz, France, at 1919 Greenwich Mean Time (GMT) on 17 July. Both crews spent about seven hours in each others' spacecraft. The two vehicles separated after 48 hours of joint flight, docked once again for practice, and then separated for a final time on 19 July at 1526 GMT. Soyuz 19 was deorbited two days later, landing near Arkalyk. The Apollo spacecraft remained in orbit until 24 July.

Both crews conducted a series of joint and separate experiments. The most important joint effort involved creating an artificial solar eclipse by maneuvering the vehicles so that the Apollo blocked the Sun and allowed the solar atmosphere,or corona, to be photo-

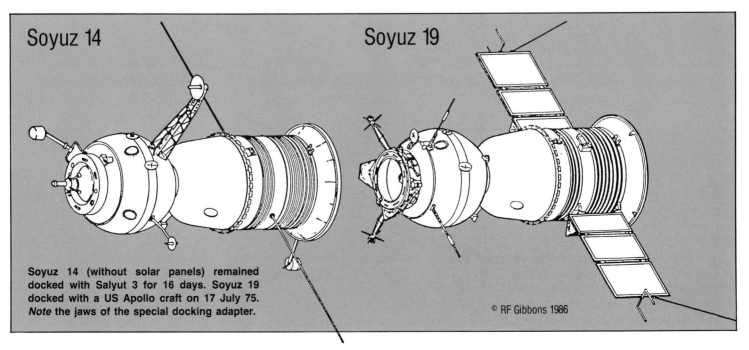

Soyuz 14 (without solar panels) remained docked with Salyut 3 for 16 days. Soyuz 19 docked with a US Apollo craft on 17 July 75. *Note* the jaws of the special docking adapter.

© RF Gibbons 1986

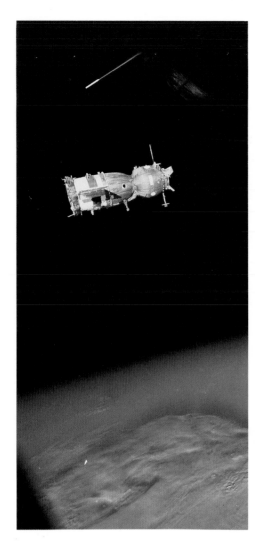

'Glad to see you'—words in English spoken by Col Alexei Leonov when he met Gen Thomas Stafford in the Apollo-Soyuz docking module. *Above:* Soyuz 19 as seen from a window of Apollo. *Right:* Soyuz 19 at rest. *Note* the personnel photos at *lower left.*

graphed from Soyuz 19. The Soviet and American crews also employed several kinds of mass spectrometers in experiments designed to measure the density of atomic oxygen and nitrogen at 225 kilometers altitude. Each spacecraft carried an identical type of fungus, cultivated prior to the mission in the US and USSR, in order to study the effects of weightlessness on its growth pattern. Hair and skin samples were taken from all five men in order to determine the level of microbial transfer that took place. Finally, the docking unit airlock contained a furnace that was used for experiments in zero-gravity metallurgy and crystal formation techniques.

Separate Soviet scientific efforts involved photography and biological experiments. The crew of Soyuz 19 photographed the Earth's horizon under varying conditions in order to gather data concerning the atmosphere. Further efforts to photograph the Sun's corona were attempted. Microorganisms, plants and fish were grown in thermostatically controlled environmental capsules called Biokats.

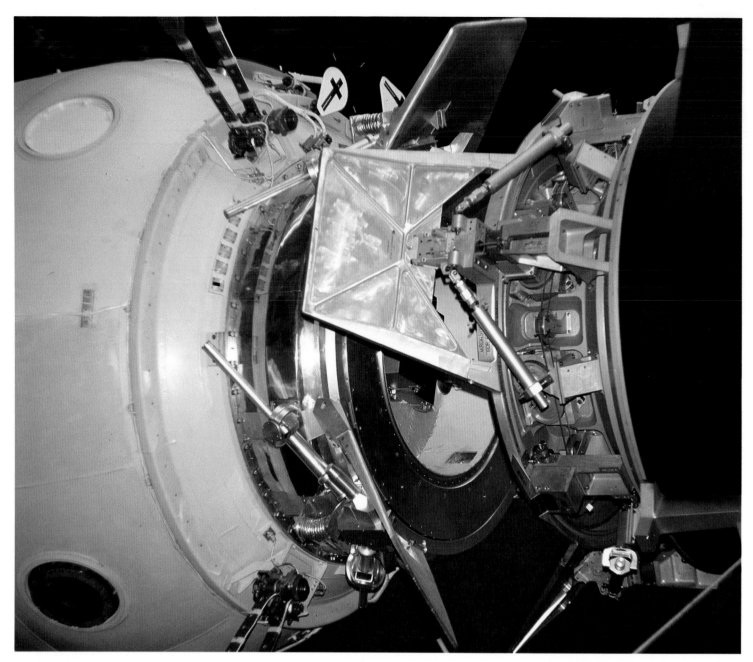

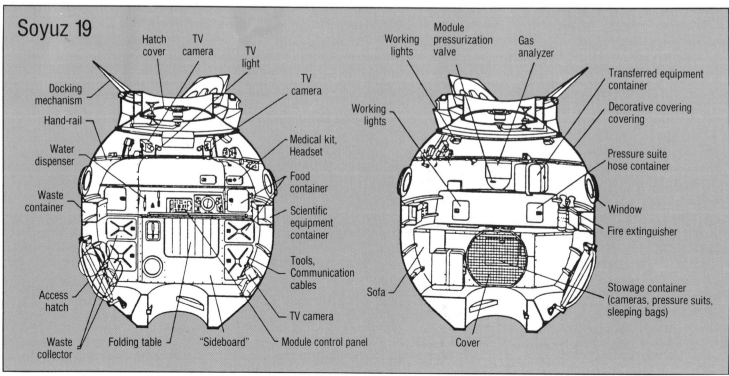

Soyuz 19

Docking mechanism

Hand-rail

Water dispenser

Waste container

Access hatch

Waste collector

Hatch cover

TV camera

TV light

TV camera

Medical kit, Headset

Food container

Scientific equipment container

Tools, Communication cables

TV camera

Folding table

"Sideboard"

Module control panel

Working lights

Module pressurization valve

Gas analyzer

Working lights

Transferred equipment container

Decorative covering covering

Pressure suite hose container

Window

Fire extinguisher

Sofa

Stowage container (cameras, pressure suits, sleeping bags)

Cover

The Soyuz 19 docking adapter *(top left)* and an illustration of the Soyuz 19 orbital module *(bottom left)* with docking adapter at top. The illustration *below* shows how this system interfaced with the docking module that was carried by the Apollo spacecraft. It is in contrast to the conventional Soyuz docking gear *(above)* which is used when docking the Soyuz spacecraft at the Salyut (or later Mir) space station.

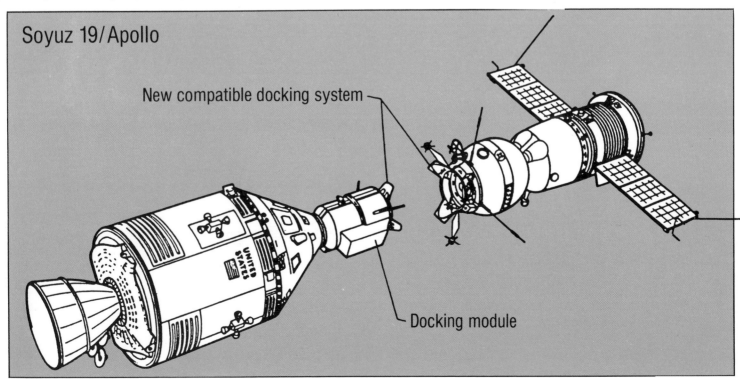

Soyuz 19/Apollo

New compatible docking system

UNITED STATES

Docking module

Kosmos 772
Launch Date: 29 September 1975
Recovery Date: 2 October 1975
Total Weight: 6800 kilograms

Kosmos 772 was probably a systems test for a three-man crew with an early Soyuz-T prototype. Although the spacecraft relied on chemical batteries for power, it stayed in space for 24 hours longer than a regular Soyuz ferry variant mission. This fact, along with Soyuz telemetry formats, suggested that the Soviets were developing a larger version of the basic Soyuz vehicle.

Soyuz 20
Launch Date: 17 November 1975
Recovery Date: 16 February 1976
Total Weight: 6570 kilograms

This flight was the second unmanned Soyuz mission and the last to dock with Salyut 4. The primary mission objective was probably to demonstrate a three-month power-down and recovery capability in anticipation of manned missions to Salyut 5. The Soviets also indicated that the vehicle was engaged in testing modifications to the Soyuz docking apparatus. The descent module was used for biological experiments. A microclimate was created for turtles, drosophila, cacti, gladioli bulbs, vegetable seeds, corn and legumes. The results of the experiments in Soyuz 20 were compared

Above: The Soyuz 19 as seen from Apollo. Not obvious here, Soyuz 19 underwent a partial color change for location by Apollo sensors.

with those of the biosat Kosmos 782, which had a similar payload of plants and animals.

Soyuz 21
Launch Date: 6 July 1976
Recovery Date: 24 August 1976
Total Weight: 6570 kilograms
Crew: B Volynov and V Zholobov

Soyuz 21 was the first flight to Salyut 5. It is widely believed in the West that this flight was terminated early, perhaps due to a malfunction in the Salyut environmental control system, because the landing took place at night some 300 kilometers away from the usual Soyuz landing area. Following the recovery, both *Izvestiya* and *Pravda* reported that the cosmonauts were suffering from 'sensory deprivation.'

Soyuz 22
Launch Date: 15 September 1976
Recovery Date: 23 September 1976
Total Weight: 6500 kilograms
Crew: V Aksenov and V Bykovskiy

The last Soyuz free-flying mission (ie not tied to a Salyut station) was conducted with the back-up spacecraft for the ASTP mission. The Soviets replaced the docking unit with an East German multispectral camera,

This photo shows cosmonauts Dzhanibekov and Gurragcha training for the Soyuz 39 mission.

Above: Flying on solar power? This Soyuz instrument module detail shows its solar panel 'wings', protruding *at top* of photo.

designated MKF-6. The camera imaged a 80×55-kilometer rectangle in four bands (two visible and two infrared) at resolutions that may have approached 10 meters. Aerial and ground photography was collected of the same areas during the mission for comparison. The MKF-6 was the first piece of foreign-built equipment to fly on a Soviet manned space mission. In addition to Earth resources photography, the crew may have used the MKF-6 to collect intelligence concerning ongoing NATO maneuvers. Soyuz 22 also continued many of the Biokat experiments with fish and plants begun during the ASTP mission.

Soyuz 23
Launch Date: 14 October 1976
Recovery Date: 16 October 1976
Total Weight: 6570 kilograms
Crew: V Rozhdestvesnkiy and V Zudov

Essentially a repetition of the Soyuz 15 docking attempt failure, Soyuz 23 apparently also suffered from a malfunction during the final automated approach sequence. Like its predecessor, the landing and recovery were also beset with problems. Soyuz 23 landed in Lake Tengiz, approximately two kilometers from shore in the midst of a blizzard with recorded temperatures as low as –20 degrees Celsius. Attempts to reach the craft with rafts were defeated by the weather, and helicopters had to be employed to tow the capsule to shore.

Kosmos 869
Launch Date: 29 November 1976
Recovery Date: 17 December 1976
Total Weight: 6800 kilograms

Kosmos 869 was a developmental flight for the Soyuz-T variant.

Soyuz 24
Launch Date: 7 February 1977
Recovery Date: 25 February 1977
Total Weight: 6570 kilograms
Crew: Y Glazkov and V Gorbatko

Soyuz 24 employed a manual docking regime in order to overcome the difficulties experienced with automated attempts. The last mission to Salyut 5 lasted only 16 days. Since Salyut 5 was a military space station, Western experts have speculated that the data gathered by the cosmonauts was considered perishable, and that this consideration prompted such a short mission. A day after Soyuz 24 left the station an unmanned reentry capsule—probably containing additional photographic intelligence—was deorbited.

Soyuz 25
Launch Date: 9 October 1977
Recovery Date: 11 October 1977
Total Weight: 6570 kilograms
Crew: V Kavalenok and V Ryumin

The first attempt to dock with the Salyut 6 second generation space station was a failure. Soyuz 25 made four attempts to complete the docking sequence, but again the automatic mode apparently malfunctioned. The crew was forced to terminate the mission because of the power supply limitations of the Soyuz ferry variant.

Soyuz 26
Launch Date: 10 December 1977
Recovery Date: 16 March 1978
Total Weight: 6570 kilograms
Crew: G Grechko and Y Romanenko

Soyuz 26 was a flight of many accomplishments. The crew broke the 84-day mission record set by the US Skylab 4 by 12

days, performed the first Soviet EVA in nine years, performed the first in-orbit refueling with a Progress tug, received two visiting crews and transferred to another Soyuz vehicle for recovery. This mission established a pattern of Soviet space station operations which continues to date. A long-duration crew occupies the space station, which receives a series of visits made by shorter-duration crews. When a short-duration crew disembarks, it usually takes the older of the two Soyuz vehicles docked at the space station, because until recently Soyuz was only reliable for 90 days in a powered-down condition. After 96 days in orbit, Grechko and Romanenko left Salyut 6 on Soyuz 27.

Soyuz 27
Launch Date: 10 January 1978
Recovery Date: 16 January 1978
Total Weight: 6570 kilograms
Crew: V Dzhanibekov and O Makarov

The first visiting crew to a Salyut station engaged in five days of joint experiments with the long-duration hosts, switched their individually contoured seats into Soyuz 26 and used that vehicle to return to Earth.

Soyuz 28
Launch Date: 2 March 1978
Recovery Date: 10 March 1978
Total Weight: 6570 kilograms
Crew: A Gubarev and V Remek

Soyuz 28 carried the first intercosmonaut, Vladimir Remek, the son of the then-Defense Minister of Czechoslovakia. Experiments conducted during the visit included materials processing, astronomical

Right: **Cosmonaut V Aksyonov works with the MKF-6 camera aboard Soyuz 22.** *Note* **the photo of an apparent family scene, behind him.**

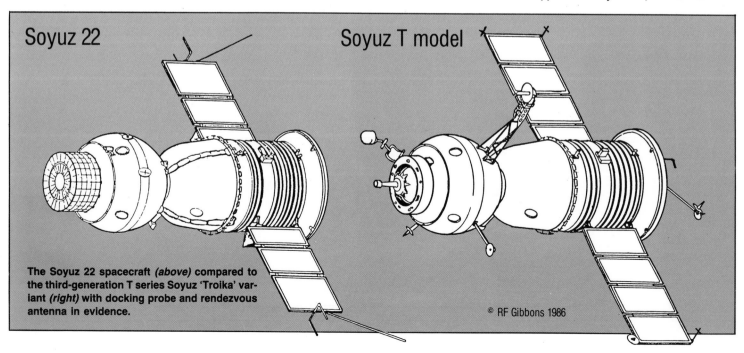

The Soyuz 22 spacecraft *(above)* **compared to the third-generation T series Soyuz 'Troika' variant** *(right)* **with docking probe and rendezvous antenna in evidence.**

Soyuz 22   Soyuz T model

© RF Gibbons 1986

observations and biomedical examinations. Gubarev and Remek returned in Soyuz 28, landing some 300 kilometers west of Tselinograd.

Kosmos 1004

Launch Date: 4 April 1978
Recovery Date: 15 April 1978
Total Weight: 6800 kilograms

Kosmos 1004 was an unmanned development flight of the Soyuz-T variant.

Soyuz 29

Launch Date: 15 June 1978
Recovery Date: 3 September 1978
Total Weight: 6570 kilograms
Crew: A Ivanchenkov and V Kovalenok

The crew of Soyuz 29 spent 140 days in orbit, moving the Soviets ahead of the US in manhours in space for the first time since 1965. The mission saw two visits by other Soyuz crews, three rendezvous with Progress resupply vehicles (including two refuelings) and an EVA. Experimental efforts concentrated on material processing, using the Splav and Kristall furnaces (see Appendix I for details). In addition the crew conducted numerous Earth-resources photography sessions and conducted various geophysical and astronomical experiments. The cosmonauts used Soyuz 31 for their return to Earth.

Soyuz 30

Launch Date: 27 June 1978
Recovery Date: 5 July 1978

Above: The Salyut space station *(left)*-Soyuz spacecraft pairing on exhibition. With these vessels, the Soviets have built an impressive record of successful missions.

Total Weight: 6570 kilograms
Crew: M Hermaszewski and P Klimuk

Soyuz 30 carried the first Polish cosmonaut, Miraslav Hermaszewski. During their seven-day visit the crew of Soyuz 30 participated in a materials processing experiment, called Sirena, that produced samples of cadmium-mercury-telluride. Multispectral photography was taken of southern Poland in conjunction with aerial photography of the same area for comparison purposes. Klimuk and Hermaszewski returned in the spacecraft that had brought them to Salyut 6.

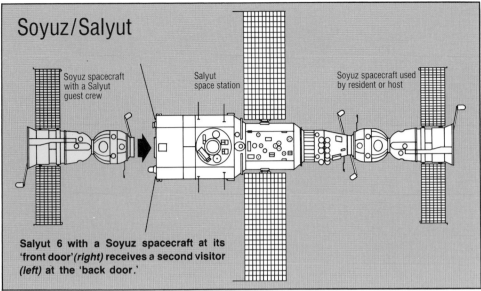

Soyuz/Salyut

Soyuz spacecraft with a Salyut guest crew

Salyut space station

Soyuz spacecraft used by resident or host

Salyut 6 with a Soyuz spacecraft at its 'front door' *(right)* **receives a second visitor** *(left)* **at the 'back door.'**

Soyuz 31

Launch Date: 26 August 1978
Recovery Date: 3 September 1978
Total Weight: 6570 kilograms
Crew: V Bykovskiy and S Jahn

The first East German cosmonaut (Sigmund Jahn) was carried into orbit by Soyuz 31. The two visiting cosmonauts from this flight also participated in materials processing and Earth resources photography. The combined crew worked on a project called Berolina, which was designed to produce ampules of bismuth-antimony and lead-telluride. Multispectral photography was carried out with two East German camera systems, the MFK-6M and the KATE-140. Bykovskiy and his East German flight en-

gineer returned to Earth in Soyuz 29, leaving a fresh craft for the main crew.

Kosmos 1074

Launch Date: 31 January 1979
Recovery Date: 1 April 1979
Total Weight: 6800 kilograms

Kosmos 1074 was the last of the unmanned Soyuz-T development flights carried out under the generic Kosmos label.

Soyuz 32

Launch Date: 25 February 1979
Recovery Date: 19 August 1979
Total Weight: 6570 kilograms
Crew: V Lyakov and V Ryumin

Soyuz 32 set yet another mission duration record, this time for 175 days in orbit. Because of an engine failure on Soyuz 33 (see entry below for details), the main crew did not receive any visitors during their stay on board Salyut 6. There were, however, three Progress missions to the station during this time. Initial efforts concentrated upon refurbishing the station, which was one and a half years old at the time. Among other tasks, a ruptured fuel tank had to be purged and then taken off-line. Materials processing continued to receive the most emphasis among the experiments, followed by Earth-resources photography, geophysical observations and astronomical measurements. This final category caused some difficulty for the Soyuz 32 crew. The last of the Progress resupply mission brought up a radio telescope called KRT-10. It was deployed outside of the aft docking port and became tangled when the crew tried to jettison it prior to departure. This situation required an EVA, which Lyakhov and Ryumin performed, despite having been in space for almost six months. Finally, there was the problem of how to return to Earth, given the

aborted Soyuz 33 flight. On 6 June the unmanned Soyuz 34 was launched and docked with Salyut 2 days later. Soyuz 32 was used to carry materials processing results back to Earth on 13 June, and the crew made their descent in Soyuz 34.

Soyuz 33

Launch Date: 10 April 1979
Recovery Date: 12 April 1979
Total Weight: 6570 kilograms
Crew: G Ivanov and N Rukavishnikov

The flight of the first Bulgarian cosmonaut (Georgiy Ivanov) suffered a major engine failure during the docking attempt with the Salyut complex. With the major engine gone and Soyuz 33 and the space station separating from each other at 100 kilometers per hour, Soviet mission control decided to use the craft's backup engine for a direct ballistic descent from Soyuz 33's 346×298-kilometer orbit. The backup engine worked, but it did not shut off automatically at the required time, and thrust had to be terminated manually. During reentry, the crew sustained G-forces over twice those of a normal recovery profile, but they emerged unscathed after touchdown some 320 kilometers southeast of Dzhezkazgan.

Soyuz 34

Launch Date: 6 June 1979
Recovery Date: 9 August 1979
Total Weight: 6570 kilograms

Soviet mission planners were reluctant to use Soyuz 32 to return cosmonauts Lyakhov and Ryumin from Salyut because the craft's engines were from the same lot as those of the ill-fated Soyuz 33 and because the craft had exceeded its 90-day rating period for powered-down operations. But the Soviets were also reluctant to risk a crew while they conducted extensive tests of the modified

The Soyuz (*left*)-Salyut pairing. The second Salyut docking port is 'important for replacing crews, carrying out rescue operations and delivering foodstuffs and equipment.'

Above: At Yuri Gagarin Cosmonaut Training Center in 'Star City,' near Moscow, cosmonauts engage in underwater repairs of a spacecraft mockup, in a 'sensory training' exercise. *Below:* Soyuz 14 commander Pavel Popovich (*at right*) briefs flight engineer Yuri Artyukhin.

engines on Soyuz 34. Thus, Soyuz 34 was launched unmanned and underwent only two days of tests before it docked with the Salyut complex and was certified as reliable for returning the crew to Earth.

Soyuz T-1
Launch Date: 16 December 1979
Recovery Date: 25 March 1980
Total Weight: 7000 kilograms

The final unmanned test of the redesigned Soyuz was given the first designator of that series. It took three days to maneuver the new craft for docking with Salyut 6, and an additional two days of free flight were conducted between undocking and recovery. With the exception of the external dimensions, most of the Soyuz's major systems had been extensively modified. The solar panels were brought back, probably as a result of the Soyuz 15, 23 and 25 docking failures that had demonstrated the danger of relying solely on chemical batteries. The interior of the descent capsule was redesigned so as to hold up to three cosmonauts in spacesuits. Advanced digital avionics were also installed in the spaceship. The rescue and reentry systems were extensively modified in order to provide greater safety for the crew. Finally, the fuel system was modified so that the main and the maneuvering engines used the same fuel, hydrazine, which is also used by Salyut and Progress. Over the next two years the Soviets gradually phased out the older generation of Soyuz craft in favor of the Soyuz-T.

Soyuz 35
Launch Date: 9 April 1980
Recovery Date: 11 October 1980
Total Weight: 6570 kilograms
Crew: L Popov and V Ryumin

The Soyuz 35 crew spent 185 days in orbit, received four visiting crews and were resupplied by four Progress craft. Around a quarter of the cosmonauts' time was spent refurbishing the Salyut 6 station, which marked its 1000th day in orbit during their tenure. Scientific work included some 70 materials processing experiments, collection of 3500 multispectral photographs of the Earth's surface and 40,000 spectral measurements of the atmosphere. Biological experiments concentrated on growing plants in zero-gravity using specially designed green-

Below: Soyuz 35 crew, Popov *(left)* and Ryumin at the Gagarin Cosmonaut Training Center. *Right:* On a monumental scale: amphitheater-like Baikonur site lends a sense of the imminent to Soyuz T-6.

houses. The crew returned to Earth in the Soyuz 37 vehicle.

Soyuz 36
Launch Date: 26 May 1980
Recovery Date: 3 June 1980
Total Weight: 6570 kilograms
Crew: B Farcas and V Kubasov

Soyuz 36 brought Bertalan Farcas, the first Hungarian cosmonaut, to the Salyut complex. In addition to the usual observation and photography of the intercosmonaut's country, the joint crew spent most of the seven-day visit using the materials processing furnaces. The experiments produced gallium-arsenide-chromium and aluminum-chromium alloys. A total of 21 joint Hungarian-Soviet experiments were conducted during the visit. Kubasov and Farcas used the Soyuz 35 vehicle for recovery.

Soyuz T-2
Launch Date: 5 June 1980
Recovery Date: 9 June 1980
Total Weight: 7000 kilograms
Crew: V Aksenov and Y Malyshev

The first manned mission in the new Soyuz vehicle was only a partial success. One of the advanced features of the T variant (the T stands for troika, or three, indicating that it is the third generation of manned Soviet spacecraft) was said to be its automatic docking capability. Upon approaching Salyut 6, however, the crew terminated the computer-controlled sequence at about 180 meters from the station and docked manually. Since the flight was essentially a systems test, Soyuz T-2 did not stay long at the Salyut complex. The vehicle used a new reentry profile by jettisoning its orbital module prior to retrorocket burn, thus saving 10 percent in fuel expenditure over the older Soyuz variants.

Soyuz 37
Launch Date: 23 July 1980
Recovery Date: 31 July 1980
Total Weight: 6570 kilograms
Crew: V Gorbatko and Pham Tuan

This flight carried the first non-Warsaw Pact intercosmonaut, Lieutenant Colonel Pham Tuan of the Socialist Republic of Vietnam. It seems hardly coincident that this mission took place during the Moscow Summer Olympic Games. The usual tetrad of materials processing experiments, Earth-resources photography, biological experiments and biomedical observations were conducted during the visit. Many photographs of Vietnam were taken for the purposes of improving the country's agriculture

and resource exploitation capabilities. Bismuth-tellurium-selenium and gallium phosphide ampules were produced in a joint Soviet-Vietnamese experiment called Halong. Gorbatko and Pham used the Soyuz 36 vehicle for recovery.

Soyuz 38
Launch Date: 18 September 1980
Recovery Date: 26 September 1980
Total Weight: 6570 kilograms
Crew: A Mendez and Y Romanenko

A second intercosmonaut from the third world arrived at the Salyut complex via Soyuz 38, the Cuban Arnaldo Tomayo Mendez. Some 20 joint experiments were conducted during the visit, including the growth of the first organic monocrystals in space using Cuban sugar. Fifty percent of the scientific work was devoted to improving the Cuban economy. Romanenko and Mendez returned to Earth in the spacecraft they had brought, indicating that the main crew would soon depart as well.

Soyuz T-3
Launch Date: 27 November 1980
Recovery Date: 10 December 1980
Total Weight: 7000 kilograms
Crew: L Kizim, O Makarov and G Streakalov

Soyuz T-3 was the first three man flight in a decade. The mission had two major objectives: to continue major systems tests for Soyuz-T and to make significant repairs to the Salyut 6 station. The crew repaired the Salyut's thermal regulation system, the onboard control complex, the telemetry system, and the refueling system. In addition, experiments were conducted involving the growth of cadmium-mercury-telluride monocrystals, plant growth in a weightless environment and holographic photography.

Soyuz T-4
Launch Date: 12 March 1981
Recovery Date: 26 May 1981
Total Weight: 7000 kilograms
Crew: V Kovalyenok and V Savinykh

This was a rather short mission by previous standards. The crew of Soyuz T-4 were in orbit for 75 days, received two visiting crews, and were resupplied once by a Progress craft. They made the return flight in their own vehicle.

Soyuz 39
Launch Date: 22 March 1981
Recovery Date: 30 March 1981
Total Weight: 6570 kilograms
Crew: V Dzhanibekov and J Gurragcha

Soyuz 39 carried the first Mongolian, Jugderdemidyn Gurragcha, into space for a week of joint experiments with Salyut 6's long-duration crew. Return to Earth was made with the vehicle which brought the visiting crew.

Soyuz 40
Launch Date: 14 May 1981
Recovery Date: 22 May 1981
Total Weight: 6570 kilograms
Crew: L Popov and D Prunariu

Soyuz 40 was the last of the second generation manned spacecraft flights and the final mission in the initial phase of the Intercosmonaut Program. The last non-Soviet cosmonaut carried on a second generation craft was Dimitry Prunariu from Romania. The crew recovered in their own craft.

Soyuz T-5
Launch Date: 13 May 1982
Recovery Date: 10 December 1982
Total Weight: 7000 kilograms
Crew: A Berzovoy and V Lebedev

Soyuz T-5 was the first manned spacecraft to dock with the new Salyut 7 space station. The crew set a new endurance record of 211 days in orbit, received two three-member visiting crews, were resupplied by four Progress resupply/space tugs and conducted an EVA. Scientific work consisted of experiments involving the familiar catagories of materials processing, Earth resources photography, astronomical/geophysical observations and biomedical studies. The cosmonauts returned home in the Soyuz T-7 spacecraft.

Soyuz T-6

Launch Date: 24 June 1982
Recovery Date: 2 July 1982
Total Weight: 7000 kilograms
Crew: J Chretian, V Dzanibekov and A Ivanchenkov

The second phase of the manned Interkosmos program began with the first visit of a representative of a Western industrial democracy to a Soviet space station. The French spationaut Jean-Loup Chretian spent the usual week aboard Salyut 7 engaged in a

Left: **The Soyuz T-5 crew** *(left to right)*, **Berezovoi and Lebedev.** *Below:* **Soyuz T-3 Cosmonauts Streakalov, Makarov and Kizim.**

program of mainly biomedical experiments. The visiting crew returned to Earth in their own Soyuz T-6 vehicle.

Soyuz T-7

Launch Date: 19 August 1982
Recovery Date: 27 August 1982
Total Weight: 7000 kilograms
Crew: L Popov, S Savitskaya and A Serebrov

This mission carried the first woman into orbit since 1963. Svetlana Savitskaya and the rest of the visiting crew departed after a week, using the Soyuz T-5 vehicle for reentry and recovery.

Soyuz T-8

Launch Date: 20 April 1983
Recovery Date: 22 April 1983
Total Weight: 7000 kilograms
Crew: A Serebrov, G Streakalov and V Titov

The fourth docking failure of the Soyuz series, Soyuz T-8 apparently suffered a radar transponder malfunction due to damage to the external radar antenna. The crew tried an optical approach, but this attempt was aborted because the cosmonauts could not determine the closure speed between the Soyuz and the Salyut complex.

Soyuz T-9

Launch Date: 27 June 1983
Recovery Date: 23 November 1983
Total Weight: 7000 kilograms
Crew: A Aleksandrov and V Lyakov

The 149-day mission of the Soyuz T-9 crew was eventful. Three major mishaps occurred during the flight. First, a micrometeorite impacted against one of the Salyut's portholes, but did not penetrate either of the viewing port's 14mm-panes. Next, one of Salyut's main oxidizer lines ruptured during a refueling sequence with a Progress spacecraft (the line was subsequently by-passed by the crew). Finally, Soyuz T-10A failed at launch (see entry below for details), leaving the crew with a Soyuz whose powered-down condition extended well beyond any experience to date. Despite these problems the crew conducted more than 350 scientific experiments, took hundreds of photographs, conducted two EVAs, received two Progress resupply/refueling missions, operated the Salyut complex for a month while docked with Kosmos 1443 (a Progress follow-on) and returned to Earth in their original spacecraft—thereby demonstrating the much longer mission life of the Soyuz third generation vehicle.

Soyuz T-9 in the first seconds of ignition. The crew spent 149 days and 10 hours in space, much of the time being aboard Salyut 7—to which they added two solar panels during two lengthy space walks. An oxidizer line rupture in space station was troublesome but, in the end, not disastrous.

Soyuz T-10A

Launch Date: 26 September 1983
Recovery Date: 26 September 1983
Total Weight: 7000 kilograms
Crew: V Titov and G Streakolov

The second attempt to reach Salyut 7 for Titov and Streakolov was even less successful than the first. Some 90 seconds before lift-off a fire developed at the base of the launch vehicle. The escape rocket system was triggered and the capsule was lifted off of the booster just seconds before it exploded on the pad. The crew underwent a brief flight at 14 G's and landed unharmed several kilometers away from the launch pad.

Soyuz 10B

Launch Date: 8 February 1984
Recovery Date: 2 October 1984
Total Weight: 7000 kilograms
Crew: O Atkov, L Kizim and V Solovyov

The first three-member long-duration crew spent 237 days in space. During this period Salyut 7 hosted two visiting crews and five Progress refueling/resupply missions. By far the most impressive accom-

Above: **Former Voskhod 2 crewmember *(left)* Leonov consults with the Soyuz 33 crew during a landing training exercise.**

plishment of the mission was the number and difficulty of the EVAs performed during the nearly eight-month flight. Six spacewalks were performed by cosmonauts Kizim and Solovyov, and one by members of the Soyuz T-12 crew (see the entry below for details). Five of the EVAs involved efforts to repair the rupture in one of the Salyut's main oxidizer lines which occurred during the Soyuz T-9 mission. The sixth EVA was undertaken in order to install additional solar panels, similar to the EVA conducted by the crew of Soyuz T-9. While the attempts to repair the breached oxidizer line apparently did not meet with complete success, the cumulative effects of the EVAs, which amounted to more time outside of a spacecraft in orbit than during the entire previous Soviet manned space experience, will provide an invaluable experiential basis for construction of large orbital complexes planned by the Soviets for the early 1990s. At least two of the spacewalks involved using a device called the Unified Manual Instrument (known by the Russian acronym URI),

which can perform cutting, welding, spraying and soldering operations.

The crew also had time to perform numerous scientific experiments and take thousands of photographs. Doctor Oleg Atkov, a Soviet cardiovascular specialist, was able to make firsthand observations concerning the biomedical effects of lengthy zero-gravity missions. At the end of the mission the Soviets announced that one of the reasons for the lengthy flight was to determine whether cosmonauts could tolerate zero gravity for the time it would take to reach Mars. Materials processing efforts concentrated on the use of the Insparitel-M apparatus, which melts various metals and plastics and coats objects with a thin layer of the melted material in a depressurized environment.

The crew returned to Earth in the Soyuz T-11 vehicle, demonstrating that the third generation craft can remain powered-down for six months.

Soyuz T-11

Launch Date: 3 April 1984
Recovery Date: 11 April 1986
Total Weight: 7000 kilograms
Crew: Y Malyshev, R Sharma and G Streakalov

With this flight Gennadiy Streakolov, a veteran of the ill-fated Soyuz T-8 and T-10A flights, finally reached Salyut 7. Along with him came Rakesh Sharma, the first Indian in space. The mission spanned the usual week of joint experiments, which concentrated on Earth resources photography of the subcontinent, materials processing efforts that used the Insparitel-M for creating a molten alloy of germanium and silver and biomedical experiments that included using Yoga techniques to combat the debilitating effects of weightlessness. The crew used the T-10 spacecraft for their return to Earth.

Soyuz T-12

Launch Date: 17 July 1984
Recovery Date: 29 July 1984
Total Weight: 7000 kilograms
Crew: V Dzhanibekov, S Savitskaya and I Volk

This mission marked the return to space of Svetlana Savitskaya, who became the first woman to walk in space on 25 July. She and Commander Dzhanibekov spent some three and a half hours in space using the URI to cut and weld titanium and stainless steel and to solder tin and lead. The device was also used to coat anodized aluminum with molten silver. The T-12 vehicle was used for reentry and recovery.

Soyuz T-13

Launch Date: 6 June 1985
Recovery Date: 25 September 1985 (Dzhanibekov and Grechko)
Total Weight: 7000 kilograms
Crew: V Dzhanibekov and V Savinykh

Soyuz T-13 was perhaps the riskiest undertaking in the history of Soviet manned space flight. The mission was to reactivate Salyut 7, which had malfunctioned seriously in early 1985, with the result that ground controllers had lost all contact with the space station. By the time Soyuz T-13 approached the complex it was frozen, due to power loss and tumbling in space. The cosmonauts entered the station and began recharging its batteries. All the water pipes in the interior of the vehicle had burst and the interior temperature was well below freezing. It was ten days before the crew could begin living in the station again. Once Salyut's power was restored, all the damaged equipment had to be replaced, and two Progress missions were flown to the station with new equipment and fuel. The crew performed an EVA to install a

Above: The Soyuz T-10B crew (T-10A aborted at launch) resting after repairing Salyut 7's damaged fuel tanks and setting a Soviet record for EVA—35 hours. *Left:* The Soyuz T-5 crew in simulation training.

third set of additional solar panels like those emplaced during the T-9 and T-10 missions. Despite their other pressing duties, the cosmonauts were able to participate in two major Earth resources projects, perform many astrophysical and geophysical observations and conduct materials processing experiments. The mission ended with the first partial crew transfer in space, as Vladimir Dzhanibekov joined Georgi Grechko, who had arrived on Soyuz T-14, for the return to Earth on Soyuz T-13.

Soyuz T-14

Launch Date: 17 September 1985
Recovery Date: 21 November 1985 (Savinykh, Vasyutin, and Volkov)
Total Weight: 7000 kilograms
Crew: G Grechko, V Vasyutin and A Volkov

The crew of Soyuz T-14 received a resupply/refueling mission from a Progress follow-on spacecraft, conducted 400 scientific experiments and took photographs of the Earth's surface that totalled some 16 million square kilometers of land and oceans. The mission had to be terminated abruptly when Commander Vasyutin became dangerously ill, reportedly suffering 104 degrees of fever. Western observers detected the first signs of this trouble when the Salyut crew switched from open to encrypted transmissions on 13 November, and eight days later Soyuz T-14 undocked and began an unannounced reentry and recovery routine. Vasyutin was hospitalized until 20 December and apparently enjoyed a complete recovery.

Left to right: cosmonauts Vladimir Dzanibe___
and Alexander Ivanchenkov, French spationa___
Jean-Loup Chretian, cosmonaut Leonid Kiz___
spationaut Patrick Baudry and cosmonaut V___
dimir Solovyov in the assembly and test hang___
at the Baikonur cosmodrome.

4 октября 1957 года
в советском союзе
осуществлен успешный запуск
первого в мире
искусственного спутника земли

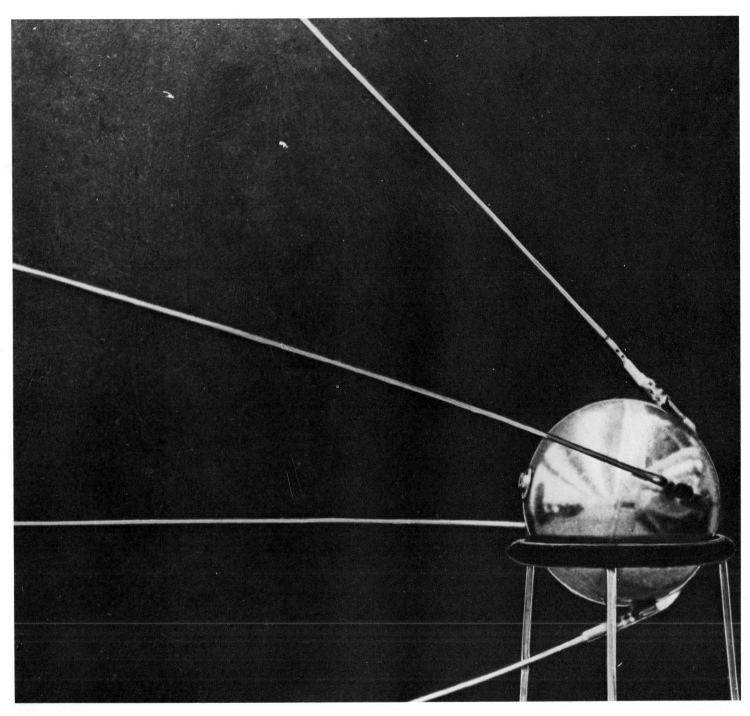

Sputnik

The most famous Soviet spacecraft remain the first three ever placed in orbit, especially Sputnik (or Traveler) 1. While there had been rumors for two years prior to the fall of 1957 that the Soviets were preparing to launch an artificial Earth satellite (the broadcast frequencies were announced in June of that year), the launch of Sputnik 1 still shocked the West. For some time attempts persisted to prove that the whole event was a fake, but the launch of Sputnik 2 silenced the critics. While the United States did manage to place two satellites in orbit after the second Sputnik, they were eclipsed by the large and complex Sputnik 3, and the space race was begun in earnest. This section describes the first three historic Soviet space launches.

Sputnik 1
Launch Vehicle: A
Launch Site: Tyuratam
Launch Date: 4 October 1957
Decay Date: 4 January 1958
Total Weight: 84 kilograms
Apogee: 947 kilometers
Perigee: 227 kilometers
Inclination: 65.1 degrees
Period: 96.2 minutes

The first artificial satellite was a spherical vehicle with four extruding rod antennas. Sputnik 1 contained no scientific instruments, but it did have chemical batteries that powered transmissions of harmonic frequencies at 20 and 40 MHz. The signal variations received by ground stations provided data concerning the Earth's ionosphere and the temperature regimes encountered by

Above: Sputnik 1, the first artificial Earth satellite. *Left:* A model of Sputnik 1 on exhibit at the Science Pavilion of the All-Union Industrial Exhibition, with data concerning both Sputnik 1 and 2.

the satellite. Orbital fluctuations and reentry profile also provided information concerning the Earth's atmosphere at the edge of space. Transmissions terminated after 21 days.

Sputnik 2
Launch Vehicle: A
Launch Site: Tyuratam
Launch Date: 3 November 1957
Decay Date: 14 April 1958
Total Weight: 508 kilograms
Apogee: 1671 kilometers
Perigee: 225 kilometers
Inclination: 65.3 degrees
Period: 103.7 minutes

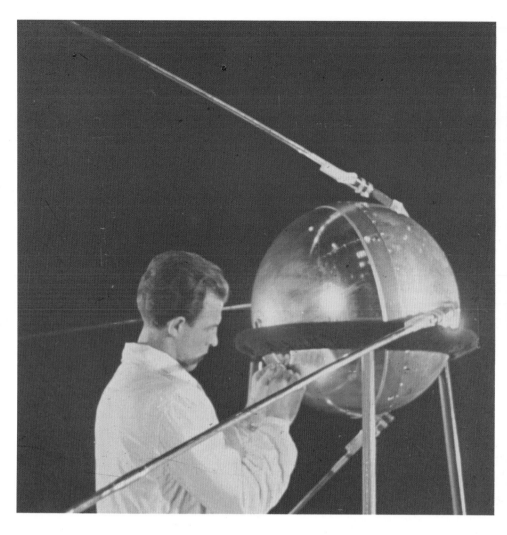

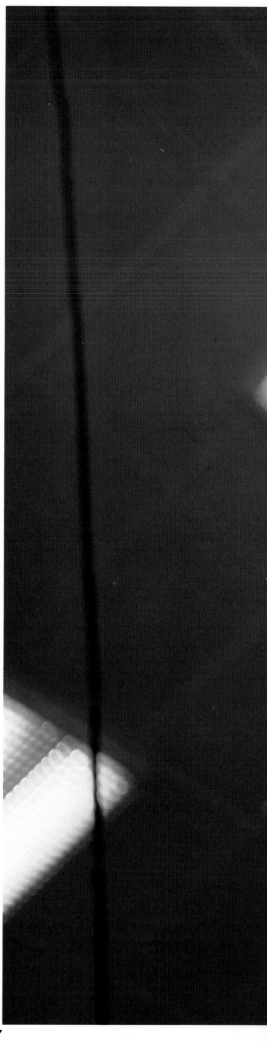

Above: 'Please comment on the use of artificial Earth satellites. What do you think could be conducted in space?' These words, asked of Soviet scientists in a questionnaire mailed by the USSR Academy of Sciences in 1955, presaged man's encounter of a new frontier. *Here,* a technician is in the seeming embrace of Sputnik 1's antennas.

Right: A closeup of Sputnik 1's antennas. 'The simplest,' as its designers called it, relayed data about the density and temperature of Earth's upper atmosphere. Sergei Korolev argued against skeptics in 1954, urging the efficacy of artificial satellites. After 'three years of crazily intensive effort,' Sputnik 1 made Korolev a visionary. *Below:* Sputnik 1's orbital paths.

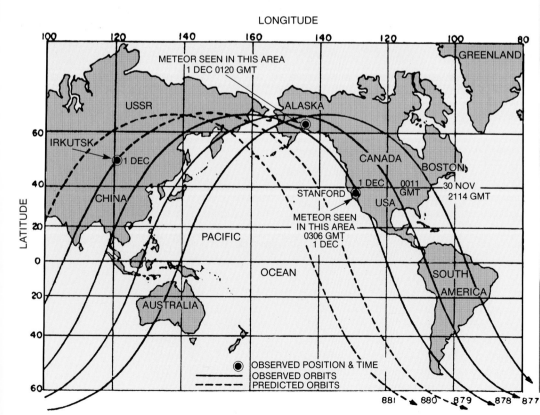

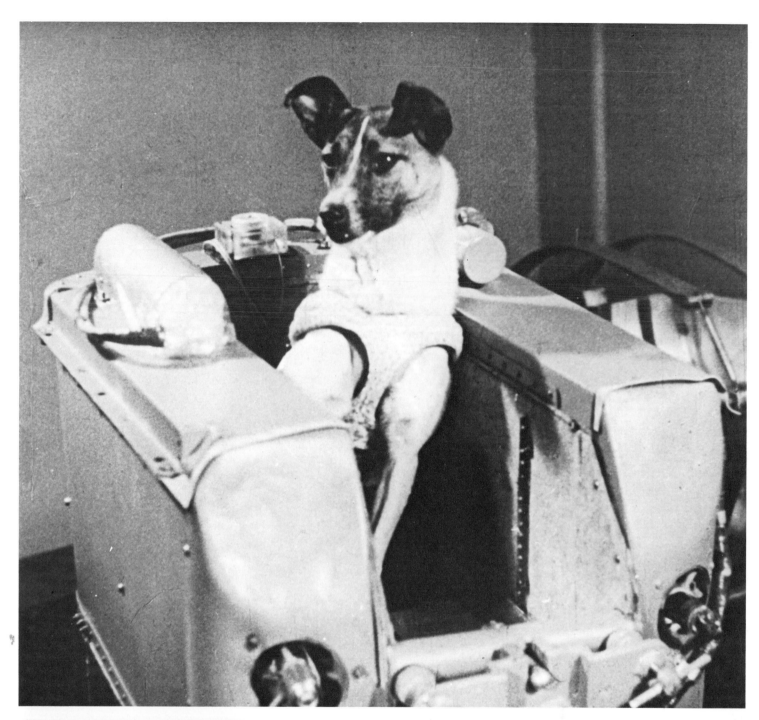

Sputnik 2 was the first biosat. The vehicle, which remained attached to the 28-meter-long rocket casing, contained the dog Layka. Data was returned on the dog's ability to adapt to the effects of spaceflight in anticipation of manned flights which were already in the planning stage. After gathering seven days of biomedical data, Layka was killed by an injection of poison, because the spacecraft could not be recovered. Other scientific instruments on the vehicle included radiation sensors and devices for measuring micrometeorite density.

Sputnik 3

Launch Vehicle: A
Launch Site: Tyuratam
Launch Date: 15 May 1958
Decay Date: 6 April 1960
Total Weight: 1327 kilograms

Above: Laika, the first space canine, and *right:* Sputnik 2, the first biosat. *At left* is the first satellite, Sputnik 1, head-on.

Apogee: 1881 kilometers
Perigee: 226 kilometers
Inclination: 65.2 degrees
Period: 106 minutes

Sputnik 3 was a conically-shaped vehicle measuring 3.57 by 1.73 meters. The vehicle was powered by solar panels embedded around the base of the main body. Various antennas protruded from the bottom, sides and top of the craft. The instrumentation package had both vacuum tubes and solid state technology. The purpose of the flight was to return geophysical data as part of the International Geophysical Year efforts. The payload continued to return data for some two years.

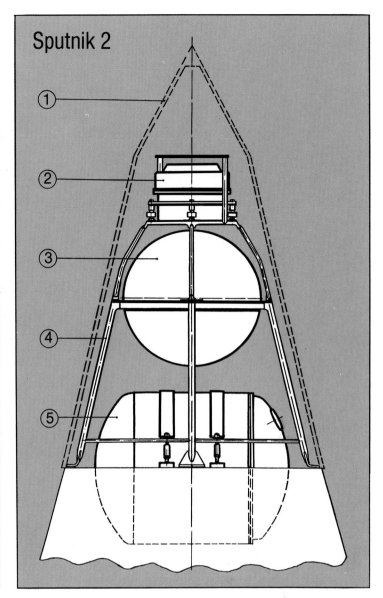

Sputnik 2

Left: Carrying six times the payload of its predecessor, Sputnik 2 was yet another Soviet 'space spectacular.' *Above:* A view of Laika in her capsule. *Right:* Sputnik 2 showing (1) nose cone, (2) solar radiation sensors, (3) additional instruments and radio transmitters, (4) instrument and equipment securing-frame and (5) sealed chamber housing Laika. Laika, the first official 'space casualty', was killed by injection because Sputnik 2 could not be recovered. On 20 August 1960, the dogs Strelka and Belka

would be the first living creatures recovered from orbital flight. *Below:* Sputnik 3, showing (1) magnetometer, (2) photo-multipliers, (3) solar batteries, (4) photon recorder, (5) ionization manometers, (6) ion traps, (7) electrostatic fluxmeters, (8) mass spectrometric tube, (9) cosmic-ray heavy nuclei recorder, (10) primary cosmic radiation sensor, (11) micrometeor recorder. *Overleaf:* Models of Sputnik 2, *foreground,* and Sputnik 3, *at rear,* at the Soviet Exhibition of Economic Achievements in 1959.

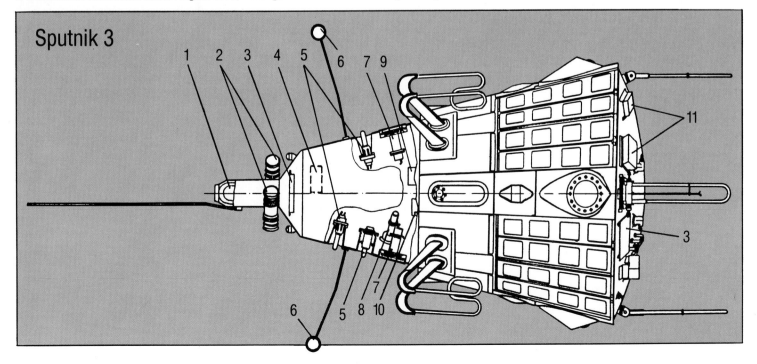

Sputnik 3

Vega

Vega, the Soviet acronym for Venus-Gallei or Venus Halley, represents the most ambitious Soviet interplanetary mission accomplished to date. The tripartite mission objective consisted in placing advanced lander modules on the surface of Venus, deploying balloons of French design in the Venusian atmosphere and, finally, using Venusian gravity to send fly-by buses to a rendezvous with Halley's comet. In addition to French and Soviet participation the program included contributions from Austria, Bulgaria, Czechoslovakia, East Germany, Hungary, Poland, the United States and West Germany. The lander modules contained a television system, a soil sample collection and analysis system, three spectrometers (mass, ultraviolet and gamma ray), a gas chromatograph, a hydrometer and an aerosol analyzer. The balloon gondolas were composed of a conical-shaped antenna that transmitted data at 1.67 GHz, a rectangular instrument package, a smaller rectangular meteorological package and a battery package. The comet probes contained a 120-kilogram instrument package for conducting experiments during the rendezvous. In addition, the comet probes were armored against hypervelocity impacts from dust particles to give them a better chance of surviving the encounter with Halley's comet long enough to return important data. The two Vega missions are discussed in this section.

Vega 1

Launch Vehicle: D-1e
Launch Site: Tyuratam
Launch Date: 15 December 1984
Fly-By Date: 11 June 1985
Landing Date: 11 June 1985

Rendezvous Date: 6 March 1986
Apogee: 204 kilometers
Perigee: 168 kilometers
Inclination: 51.6 degrees
Period: 88.2 degrees

Vega 1's lander touched down at 7 degrees 11 minutes north latitude/177 degrees 48 minutes longitude. The timing devices on the probe malfunctioned during the descent, resulting in a premature deployment of the drilling apparatus. The lander transmitted data from the Venusian surface for 21 minutes. Vega 1's balloon deployed at an altitude of 54 kilometers and returned data to Earth for 46 hours. The balloon was tracked by 20 ground stations in Australia, Sweden, the UK, US, and USSR. Vega 1's fly-by bus was the first of five probes to encounter Halley's Comet, passing within 10,000 kilometers of its nucleus on 9 March 1986.

Vega 2

Launch Vehicle: D-1e
Launch Site: Tyuratam
Launch Date: 21 December 1984
Fly-By Date: 15 June 1985
Landing Date: 15 June 1985
Rendezvous Date: 9 March 1986
Apogee: 207 kilometers
Perigee: 168 kilometers
Inclination: 52 degrees
Period: 88.2 degrees

The Vega 2 descent module fared better than its predecessor and was able to collect and chemically analyze a sample of Venusian soil from its landing site at 6 degrees 27 minutes south latitude/181 degrees 5 minutes longitude. The soil was found to be similar to that of the lunar highlands, which was analyzed during the Soviet Luna program. The balloon deployed by Vega 2 performed

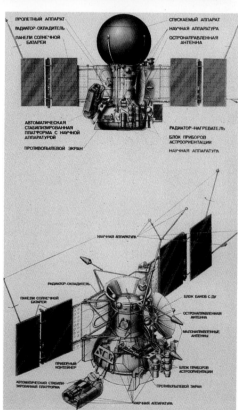

Top: A Vega mission international planning meeting. *Above:* Cutaway views show equipment to be deployed in the Vega Venus-Halley's Comet missions, which were flown by identical craft. *Right:* Scientists prepare a Vega spacecraft at the Baikonur Cosmodrome.

almost identically to the first one, lasting some 46 hours before ceasing to return data. The Vega 2 fly-by bus rendezvoused with Comet Halley three days after Vega 1, closing to within 7000 kilometers of the nucleus.

Venera

The Soviet Venus program stands in sharp contrast to the rather dismal record of the Mars program. While

Venera has had its share of failures, it has also produced dramatic successes—including the first pictures of the Venusian surface. As of 1985, the Soviets had made seven successful soft landings on Venus, placed two scientific payloads into Venusian orbit and mapped the planet from two orbital radar satellites. This section describes the Venera series and associated launches given the Kosmos designation. Venus missions carried out under the Zond probe designation are treated under that heading in this volume.

Venera (announced as Tyazheily Sputnik 4)

Launch Vehicle: A-1
Launch Site: Tyuratam
Launch Date: 4 February 1961
Decay Date: 26 February 1961
Total Weight: 644 kilograms
Apogee: 328 kilometers
Perigee: 224 kilometers
Inclination: 65 degrees
Period: 89.8 minutes

This mission was placed in near-Earth orbit during the Venus launch window, but the planetary stage failed and the vehicle burned up on reentry.

Venera 1

Launch Vehicle: A-1
Launch Site: Tyuratam
Launch Date: 12 February 1961
Fly-By Date: 19 May 1961
Total Weight: 644 kilograms
Apogee: 282 kilometers
Perigee: 229 kilometers
Inclination: 65 degrees
Period: 89.6 minutes
Aphelion: 1.019 astronomical units
Perihelion: 0.718 astronomical units
Inclination: 0.58 degrees
Period: 300 days

The first planetary probe to escape Earth orbit, Venera 1 returned data until 27 February, when it was some 7.25 million kilometers from Earth. The vehicle passed within 100,000 kilometers of Venus in mid-May.

Venera (unacknowledged)

Launch Vehicle: A-2e
Launch Site: Tyuratam
Launch Date: 25 August 1962
Decay Date: 28 August 1962
Total Weight: 890 kilograms
Apogee: 252 kilometers
Perigee: 173 kilometers
Inclination: 64.9 degrees
Period: 88.7 minutes

Despite a shift to the A-2e booster, allowing larger payloads, 1962 was a bad year for the Soviet Venus program. Three probes were orbited, but none managed to escape Earth orbit. The following two unacknowledged failures were identical to this mission.

Venera (unacknowledged)

Launch Vehicle: A-2e
Launch Site: Tyuratam
Launch Date: 1 September 1962
Decay Date: 6 September 1962
Total Weight: 890 kilograms
Apogee: 310 kilometers
Perigee: 180 kilometers
Inclination: 64.9 degrees
Period: 89.4 minutes

Venera (unacknowledged)

Launch Vehicle: A-2e
Launch Site: Tyuratam
Launch Date: 12 September 1962
Decay Date: 17 September 1962
Total Weight: 890 kilograms
Apogee: 213 kilometers
Perigee: 186 kilometers
Inclination: 64.9 degrees
Period: 88.5 minutes

Kosmos 27

Launch Vehicle: A-2e
Launch Site: Tyuratam
Launch Date: 27 March 1964
Decay Date: 29 March 1964
Total Weight: 890 kilometers
Apogee: 194 kilometers
Perigee: 181 kilometers
Inclination: 64.8 degrees
Period: 88.2 minutes

Launched during the 1964 Venus window, this mission managed a payload separation but the upper stage did not fire, stranding the vehicle in low-Earth orbit, from which it subsequently decayed. This failure was at least acknowledged to the extent of being given a Kosmos label.

Venera 2

Launch Vehicle: A-2e
Launch Site: Tyuratam
Launch Date: 12 November 1965
Fly-By Date: 27 February 1966
Total Weight: 963 kilograms
Apogee: 216 kilometers
Perigee: 203 kilometers
Inclination: 51.9 degrees

Period: 88.7 minutes
Aphelion: 1.197 astronomical units
Perihelion: 0.716 astronomical units
Inclination: 4.29 degrees
Period: 341 days

The Venera 2 and 3 spacecraft were larger versions of the Venera 1 vehicle. The main body was a cylinder with the mid-course correction engine at the aft end and the spherical payload module at the forward end. The control instruments and fuel tanks were contained in the main body between the engine and the payload. Mounted along the main body were two navigation sensors (one solar and one stellar) and a magnetometer. The other side held solar panels, thermal control radiators and a high-gain directional antenna. Venera 2 passed Venus by 24,000 kilometers at the closest approach, but it had ceased to return data by that point.

Venera 3

Launch Vehicle: A-2e
Launch Site: Tyuratam
Launch Date: 16 November 1965
Impact Date: 27 February 1966
Total Weight: 960 kilograms
Apogee: 293 kilometers
Perigee: 213 kilometers
Inclination: 51.9 degrees
Period: 89.5 minutes
Aphelion: 1.11 astronomical units
Perihelion: 0.7 astronomical units
Inclination: 4.29 minutes
Period: 316 days

The first human artifact to reach the surface of another planet, Venera 3 impacted the surface of Venus some 450 kilometers from its visible center. Unfortunately, like its immediate predecessor, Venera 3 ceased transmission prior to entering the Venusian atmosphere and no data was returned to Earth.

Kosmos 96

Launch Vehicle: A-2e
Launch Site: Tyuratam
Launch Date: 23 November 1965
Decay Date: 9 December 1965
Total Weight: 960 kilograms
Apogee: 310 kilometers
Perigee: 227 kilometers
Inclination: 51.9 degrees
Period: 89.6 minutes

The third attempt to reach Venus in 1965 failed to leave Earth orbit and was given the generic Kosmos epitaph.

Venera 4

Launch Vehicle: A-2e
Launch Site: Tyuratam

Launch Date: 12 June 1967
Entry Date: 18 October 1967
Total Weight: 1106 kilograms
Apogee: 188 kilometers
Perigee: 162 kilometers
Inclination: 51.8 degrees
Period: 87.9 degrees
Aphelion: 1.11 astronomical units
Perihelion: 0.7 astronomical units
Inclination: .4 degrees
Period: 316 days

Venera 4 through 8 were a further evolution of the basic planetary vehicle. The ma-

Opposite: The failed Venera 1 spacecraft. Above: The Venera 4 descent capsule (at bottom) was attached with metal straps.

jor difference was a heavier payload capsule, which was designed to enter the Venusian atmosphere, perform aerodynamic braking maneuvers, deploy a heat resistant parachute and descend to the surface while making atmospheric measurements. Venera 4, 5 and 6 were not designed to transmit after they reached the surface, while Venera 7 and 8 were.

The main body also contained instruments for atmospheric measurements during its

close fly-by. The Venera 4 main vehicle carried hydrogen and oxygen sensors, a magnetometer, charged particle traps and cosmic ray counters. The 383-kilogram descent module contained 11 gas analyzers, two thermometers, a barometer, a radio altimeter and an atmospheric densitometer. The payload returned data for 96 minutes during its descent, ceasing transmission at an altitude around 25 kilometers. Both components of Venera 4 gather significant data concerning Venusian atmospheric content, pressure and temperature, as well as Venus's magnetic field and hydrogen corona.

Kosmos 167
Launch Vehicle: A-2e
Launch Site: Tyuratam
Launch Date: 17 June 1967
Decay Date: 25 June 1967
Total Weight: 1106 kilograms
Apogee: 286 kilometers
Perigee: 201 kilometers
Inclination: 51.8 degrees
Period: 89.2 degrees

Kosmos 167 achieved orbit and payload separation, but the planetary stage failed to ignite.

Venera 5
Launch Vehicle: A-2e
Launch Site: Tyuratam
Launch Date: 5 January 1969
Entry Date: 16 May 1969
Total Weight: 1130 kilograms

Above: The Venera 4 descent capsule survived entry deceleration of up to 300g's, and transmitted data to within 15.5 miles of the Venusian surface. Mariner 5, launched 14 June 1967, was a near-companion to Venera 4. *Below:* Venera 4, showing (1) orbital compartment, (2) astro-orientation transducer, (3) permanent solar orientation transducer, (4) gas containers, (5) sun-earth orientation transducer, (6) magnetometer transducer, (7) narrow-focus parabolic antenna, (8) general-focus antenna, (9) temperature radiation system regulator, (10) solar battery panel, (11) correction engine unit, (12) astro-orientation micro-engines, (13) particle counter, (14) descent apparatus. Venera 4 transmitted data for 94 minutes plus.

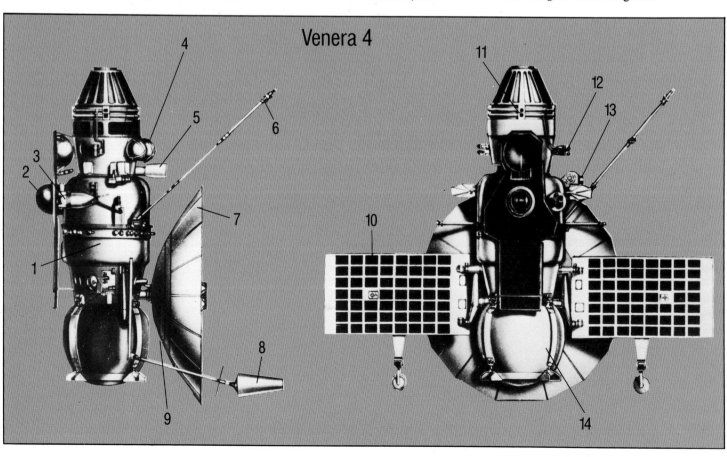

The descent module of the improved Venera 7 (above) survived extreme pressure and heat to transmit data from the Venusian surface.

Apogee: 225 kilometers
Perigee: 186 kilometers
Inclination: 51.8 degrees
Period: 88.6 minutes
Aphelion: 1.08 astronomical units
Perihelion: 0.72 astronomical units
Inclination: 2 degrees
Period: 313 days

Modifications were made to the descent module for the 1969 Venus launch window. The capsule was given better thermal and pressure shielding, thus allowing a reduction of two-thirds in the parachute size, which in turn facilitated a faster descent and closer approach to the Venusian surface. Venera 5 returned data for 53 minutes, eventually landing on Venus' night side.

Venera 6
Launch Vehicle: A-2e
Launch Site: Tyuratam
Launch Date: 10 January 1969
Entry Date: 17 May 1969
Total Weight: 1,130 kilograms
Apogee: 207 kilometers
Perigee: 186 kilometers
Period: 88.6 minutes
Aphelion: 1.08 astronomical units
Perihelion: 0.72 astronomical units
Inclination: 2 degrees
Period: 313 days

Venera 6's descent capsule returned data for 51 minutes the day following Venera 5's descent, allowing a close comparison of the two collection efforts.

Venera 7
Launch Vehicle: A-2e
Launch Site: Tyuratam
Launch Date: 17 August 1970

Far left: The Venera 8, being prepared. *Left:* A model of Venera 11/12 on exhibit. *Bottom left:* A Venera 9 lander mockup.

Landing Date: 15 December 1970
Total Weight: 1180 kilograms
Apogee: 233 kilometers
Perigee: 174 kilometers
Inclination: 51.7 degrees
Period: 88.5 minutes
Aphelion: 1.08 astronomical units
Perihelion: 0.72 astronomical units
Inclination: 2 degrees
Period: 313 days

For the 1970 launch window, the descent module was strengthened to allow transmission from the surface of Venus. No holes were drilled in the capsule hull, and the sensors deployed only after the parachute mechanism was triggered. Venera 7 transmitted for 35 minutes during the descent and for 23 minutes from the planet's surface, another Soviet space first. Surface temperatures on Venus were found to be around 475 degrees Celsius, and the atmospheric pressure to be approximately 90 times that of Earth.

Kosmos 359

Launch Vehicle: A-2e
Launch Site: Tyuratam
Launch Date: 22 August 1970
Decay Date: 6 November 1970
Total Weight: 1180 kilograms
Apogee: 910 kilometers
Perigee: 210 kilometers
Inclination: 51.5 degrees
Period: 95.5 minutes

This failure resulted from a suboptimal upper stage burn which placed the planetary vehicle in an elliptical orbit, but did not allow escape into interplanetary space. Decay and destruction followed in a few months.

Venera 8

Launch Vehicle: A-2e
Launch Site: Tyuratam
Launch Date: 27 March 1972
Landing Date: 22 July 1972
Total Weight: 1180 kilograms
Apogee: 241 kilometers
Perigee: 194 kilometers
Inclination: 51.8 degrees
Period: 88.6 days
Aphelion: 1.08 astronomical units
Perihelion: 0.72 astronomical units
Inclination: 2 degrees
Period: 313 days

Modifications to the Venera 8 mission included a soil classification device and a decision to land the vehicle on the planet's day

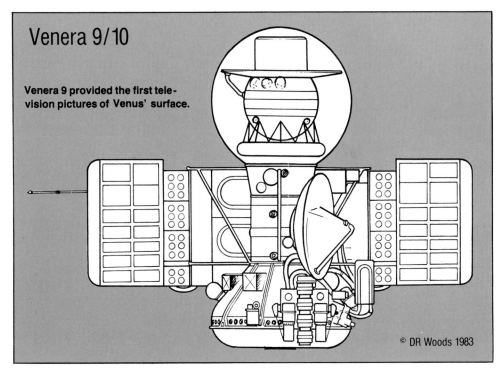

Venera 9/10

Venera 9 provided the first television pictures of Venus' surface.

© DR Woods 1983

side (all other landings had been on the night side). Venera 8 returned data for 50 minutes after touchdown, confirming the temperature and pressure readings taken by Venera 7. The Venusian soil in the vehicle's landing area resembled granite in chemical composition and had a density of 1.5 grams per cubic centimeter.

Kosmos 482

Launch Vehicle: A-2e
Launch Site: Tyuratam
Launch Date: 31 March 1972
Total Weight: 1180 kilometers
Apogee: 9813 kilometers
Perigee: 210 kilometers
Inclination: 52 degrees
Period: 201.4 minutes

For the second launch window in a row the planetary stage of a Venera mission misfired, this time sending the vehicle into a highly elliptical orbit.

Venera 9

Launch Vehicle: D-1e
Launch Site: Tyuratam
Launch Date: 8 June 1975
Insertion Date: 22 October 1975
Landing Date: 22 October 1975
Total Weight: 4936 kilograms
Apogee: 196 kilometers
Perigee: 171 kilometers
Inclination: 51.5 degrees
Period: 88.1 minutes
Aphelion: 1.11 astronomical units
Perihelion: .7 astronomical units
Inclination: .23 days
Period: 316 days
Apoapsis: 112,200 kilometers (7)
Periapsis: 1510 kilometers (7)

Inclination: 34.1 degrees
Period: 2898 minutes

The switch to the D-1e booster allowed a major increase in the payload weight which could be sent to Venus for the 1975 launch window. Venera 9 and 10 were orbiter/lander vehicles. The orbiter bus contained the flight control instrumentation, midcourse correction engine, solar panels and a scientific instrument package. The orbiter's mission was to conduct observations of the Venusian cloud tops, make measurements of Venus' upper atmosphere and conduct solar wind experiments. The following scientific equipment was contained in the orbiter module:

· four types of spectrometer
· two photometers
· a photopolarimeter
· an infrared radiometer
· a magnetometer
· charged particle traps
· a panoramic camera

The lander was encased in a double hemispheric capsule and was approximately two meters in height. It had a special cooling system to prolong instrument life on the hot Venusian surface. The primary instrument aboard the lander was a panoramic television system. TV pictures from the surface were relayed back to Earth on a boosted signal from the orbiters. Additional scientific instruments aboard the landers included three photometers, two spectrometers, accelerometers, an anemometer, a radiation densitometer and other radiation sensors.

Venera 9 began sending TV panoramas within 115 minutes of landing. The surface was brighter than had been predicted, obviating the need for the launcher's flood lights. The images revealed that the local geology was not dormant as expected, but indicative of a 'young' active planet.

Venera 10

Launch Vehicle: D-1e
Launch Site: Tyuratam
Launch Date: 14 June 1975
Insertion Date: 25 October 1975
Landing Date: 25 October 1975
Total Weight: 5033 kilograms
Apogee: 206 kilometers
Perigee: 162 kilometers
Inclination: 51.5 degrees
Period: 88.1 minutes
Aphelion: 1.11 astronomical units
Inclination: .23 degrees
Period: 316 days
Apoapsis: 113,900 kilometers
Periapsis: 1620 kilometers
Inclination: 29.5 degrees
Period: 2963 minutes

The Venera 10 lander touched down some 2200 kilometers from Venera 9. The vehicle apparently landed in a field of congealed lava. During its 65 minutes of operation on the surface, Venera 10 returned data on wind velocity, surface temperature, and rock density.

Venera 11

Launch Vehicle: D-1e
Launch Site: Tyuratam
Launch Date: 9 September 1978
Fly-By Date: 25 December 1978
Landing Date: 25 December 1978
Total Weight: 4940 kilograms
Apogee: 205 kilometers
Perigee: 170 kilometers
Inclination: 51.6 degrees
Period: 88.2 minutes
Aphelion: 1.11 astronomical units
Perihelion: .7 astronomical units
Inclination: 2.3 degrees
Period: 316 days

The 1978 Venus shots were lighter than Venera 9 and 10, because the energy requirements for reaching Venus were much higher. In addition, the non-lander portion of the Venera 11 and 12 vehicles did not go into orbit around Venus but conducted a fly-by at around 35,000 kilometers from the Venusian surface. The fly-by buses were designed to gather data in interplanetary space before and after the landers were sent to the surface of Venus. The instruments aboard the fly-by buses collected information on cosmic gamma ray flares, high energy particles, interplanetary plasma and the solar wind. Besides Soviet equipment, the Venera fly-by

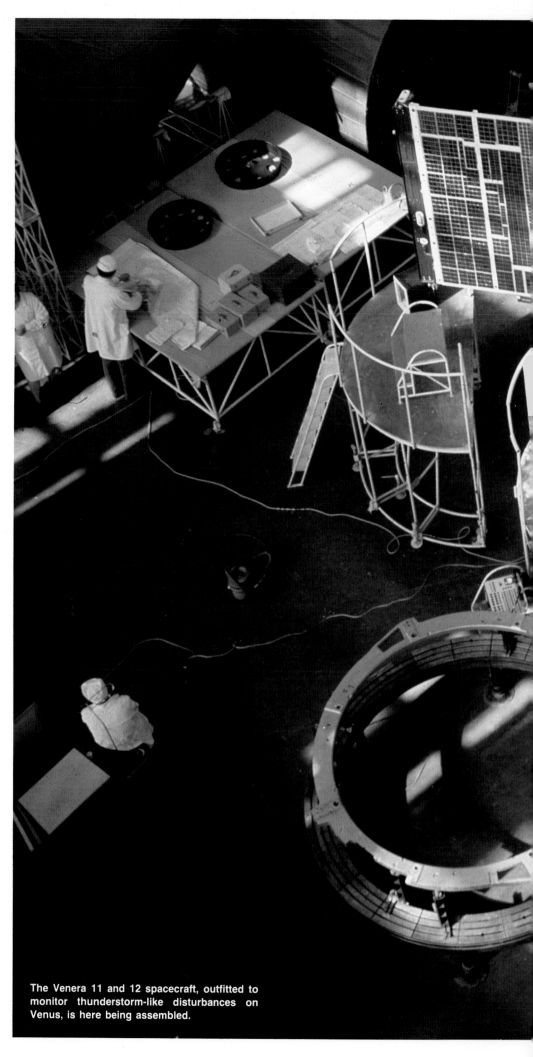

The Venera 11 and 12 spacecraft, outfitted to monitor thunderstorm-like disturbances on Venus, is here being assembled.

buses also carried the French-built SNEG 2MZ omnidirectional gamma ray detector.

The Venera 11 lander touched down at 0624 Moscow time on Christmas Day. During the descent the payload conducted a series of measurements concerning electrical activity in, and chemical composition of, the Venusian atmosphere. The lander operated on the surface for 110 minutes transmitting temperature and atmospheric pressure data back to Earth via the fly-by bus, but the lander's TV system apparently failed to function.

Venera 12

Launch Vehicle: D-1e
Launch Site: Tyuratam
Launch Date: 14 September 1978
Fly-By Date: 21 December 1978
Landing Date: 21 December 1978
Total Weight: 4940 kilograms
Apogee: 207 kilometers
Perigee: 164 kilometers
Inclination: 51.5 degrees
Period: 88.2 minutes
Aphelion: 1.11 astronomical units
Perihelion: .7 astronomical units
Inclination: 2.3 degrees
Period: 316 days

Venera 12's lander was actually the first to reach the surface, landing at a spot some 800 kilometers from where the Venera 11 lander touched down four days later. The Venera 12 vehicle remained viable for 95 minutes on the surface, but it too failed to return any TV pictures.

Venera 13

Launch Vehicle: D-1e
Launch Site: Tyuratam
Launch Date: 30 October 1981
Fly-By Date: 1 March 1982
Landing Date: 1 March 1982
Apogee: 212 kilometers
Perigee: 167 kilometers
Inclination: 51.6 degrees
Period: 88.3 minutes
Aphelion: 1.11 astronomical units
Perihelion: .7 astronomical units
Inclination: 2.3 degrees
Period: 316 days

The 1981 Venus probes were also fly-by/lander combinations. The fly-by buses again conducted studies of gamma ray flares before and after lander separation. The interplanetary studies payloads included an Austrian-built magnetometer.

During the descent, the landers employed the following instruments for atmospheric measurements:

· 3 spectrometers (mass, x-ray and optical)
· a hydrometer
· a nephelometer
· a gas chromatograph

The landers returned the first color television pictures of the Venusian surface and sky, collected and conducted chemical analysis of soil samples and took seismic measurements. The Venera 13 lander touched down at 7 degrees 30 minutes south latitude/303 degrees longitude. Both landing sites were chosen from radar maps made by the US Pioneer-Venus spacecraft in the hopes of placing the probes in the area of volcanic activity.

Venera 14

Launch Vehicle: D-1e
Launch Site: Tyuratam
Launch Date: 4 November 1981
Fly-By Date: 5 March 1982
Landing Date: 5 March 1982
Apogee: 211 kilometers
Perigee: 169 kilometers
Inclination: 51.6 degrees
Period: 89.7 minutes
Aphelion: 1.11 astronomical units
Perihelion: .7 astronomical units
Inclination: 2.3 degrees
Period: 316 days

The Venera descent module landed at 13 degrees 15 minutes south latitude/310 degrees 9 minutes longitude. The probe transmitted data via its fly-by bus for 57 minutes.

Venera 15

Launch Vehicle: D-1e
Launch Site: Tyuratam
Launch Date: 2 June 1983
Insertion Date: 10 October 1983
Apogee: 199 kilometers
Perigee: 157 kilometers
Inclination: 51.6 degrees
Period: 88.1 minutes
Aphelion: 1.11 astronomical units
Perihelion: .7 astronomical units
Inclination: 2.3 degrees
Period: 316 days
Apoapsis: 65,000 kilometers
Periapsis: 1000 kilometers
Inclination: 87 degrees
Period: 24 hours

The basic vehicle design employed since Venera 9 was extensively modified for the 1983 Venus launch window. Venera 15 and 16 were orbiters equipped with side-looking radars used for mapping the surface of

Venus. The radar, Polyus-V, replaced the lander module on the vehicle. Polyus-V had a 6×1.4-meter antenna with a twenty-minute scanning period, an overall imaging area of one million square kilometers and a resolution of one to two kilometers. The Soviets eventually produced maps covering some 120 million square kilometers of the northern Venusian hemisphere. Venera 15 ceased operations in March 1985.

The orbiter also carried an infrared spectrometer-interferometer of East German design for measuring atmospheric temperatures. The two orbiters eventually compiled a temperature map of the northern hemisphere of Venus. In keeping with practice of earlier Venera missions, the radar orbiters collected data on charged particles and cosmic rays en route to Venus.

Venera 16

Launch Vehicle: D-1e
Launch Site: Tyuratam
Launch Date: 6 June 1983
Insertion Date: 14 October 1983
Apogee: 199 kilometers
Perigee: 159 kilometers
Inclination: 51.5 degrees
Period: 88.1 minutes
Aphelion: 1.11 astronomical units
Perihelion: .7 astronomical units
Inclination: 2.3 degrees
Period: 316 days
Apoapsis: 65,000 kilometers
Periapsis: 1000 kilometers
Inclination: 87 degrees
Period: 24 hours

The Venera 16 mission is essentially identical to that of Venera 15. The Venera 16 radar was still operational in late 1985.

Vertikal

As part of the Interkosmos Program the Soviet Union conducted eight suborbital launches between 1970 and 1979 under the Vertikal designator. The launches were conducted from Kapustin Yar using the first stage of the B-1 booster. Contributions to payload instrumentation and launch assistance have been made by Bulgaria, Czechoslovakia, East Germany, Hungary, Poland and Romania.

The general Vertikal flight profile involved an ascent to an altitude between 500 and 1500 kilometers. The payload usually separated early in the launch sequence, and conducted measurements and experiments during its suborbital trajectory. The payload

Right: Radar replaced the landing module on Venera 15, which compiled radar and infrared maps of the northern Venusian hemisphere during the 1978 'invasion' of Venus.

The Soviets have conducted suborbital geophysical projects with Vertikal sounding rockets such as this one on the launchpad.

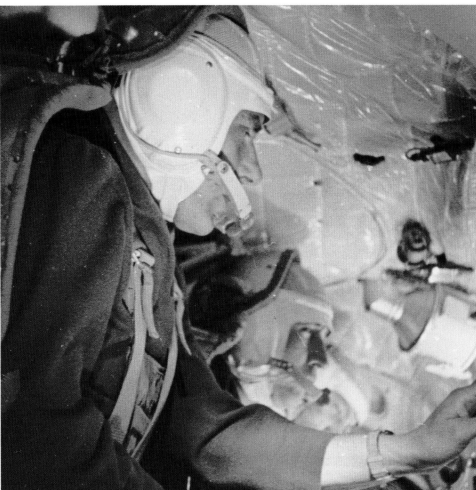

was then recovered via parachute. Experiments conducted during the Vertikal program were similar to those geophysical investigations carried out under the auspices of the orbital Interkosmos Program (see the entry in this volume for details).

Vesta

V esta is the name the Soviets have given to a future joint Franco-Soviet probe that would visit both Mars and an asteroid sometime in the 1990s. Vesta will probably be a modification of the Venera/Vega design. Mission planning calls for a Mars fly-by, after which the Vesta craft would deploy a separate asteroid probe of French design. Plans call for placing an instrumentation package containing magnetometers, penetrometers, radiometers, spectrometers and TV cameras on the surface of an asteroid. The estimated five-year mission would constitute the longest Soviet interplanetary flight to date.

Voskhod

B etween the first generation manned Vostok flights and the beginning of the second generation Soyuz program, the Soviets conducted two manned

At left: Vertikal. *Above:* The crew of Voskhod 1: Komarov, Feoktistov and Yegorov.

missions under the Voskhod, or Sunrise, label and several related unmanned Kosmos launches. The Voskhod flights made history by carrying the first multi-man crew into space and hosting the first EVA. The Voskhod missions and their related Kosmos flights are described below.

Kosmos 47

Launch Vehicle: A-2
Launch Site: Tyuratam
Launch Date: 6 October 1964
Recovery Date: 7 October 1964
Total Weight: 5320 kilograms
Apogee: 413 kilometers
Perigee: 177 kilometers
Inclination: 64.8 degrees
Period: 90 minutes

Kosmos 47 preceded Voskhod 1 by only six days and flew an almost identical orbit. The mission was probably meant to test both the modifications to the basic Vostok design and the A-2 launch vehicle for manned flight.

Voskhod 1

Launch Vehicle: A-2
Launch Site: Tyuratam
Launch Date: 12 October 1964

Recovery Date: 13 October 1964
Total Weight: 5320 kilograms
Apogee: 409 kilometers
Perigee: 178 kilometers
Inclination: 65 degrees
Period: 90.1 minutes
Crew: K Feoktistov, V Komarov and B Yegorov

The introduction of the A-2 booster for manned space flight allowed an increase in payload weight from the Vostok level of 4700 kilograms to 5320 kilograms. The Soviets used the extra weight-to-orbit capability for modifications to the Vostok vehicle (see the Vostok entry for details). On top of the spherical reentry vehicle the Soviets mounted a reserve retrorocket, which, unlike the primary liquid fuel engine contained in the instrument module, was a solid rocket motor capable of generating 12,000 kilograms of thrust.

The interior of the reentry capsule was modified to accommodate up to three cosmonauts by removing the ejection seat mechanism. This was possible because the Soviets had developed retrorocket technology to the point where a soft landing could be assured for the crew by firing the engines just before touchdown, thus obviating the need for ejection and separate descent *a la* Vostok. In addition, Soviet confidence in life support

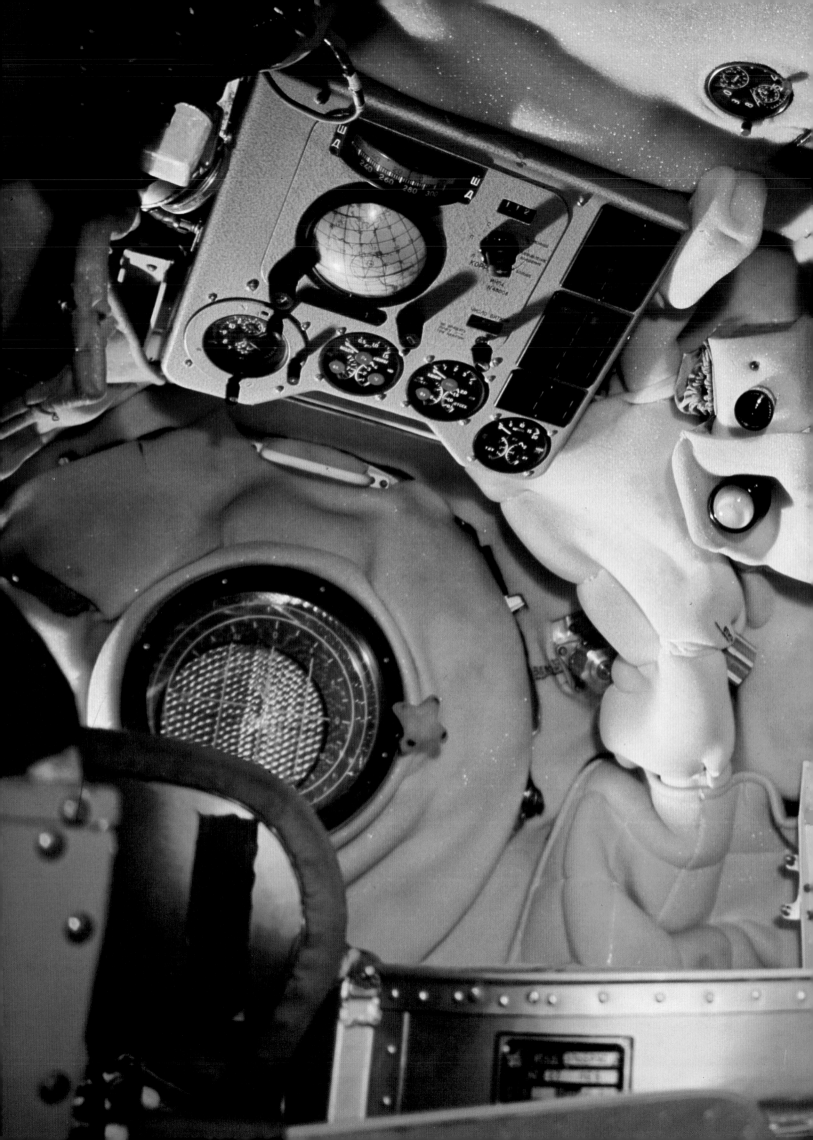

This is a view of the rather cozy interior of the Voskhod 1, designed to carry out the first multi-man space flight. The three-man crew did not wear spacesuits.

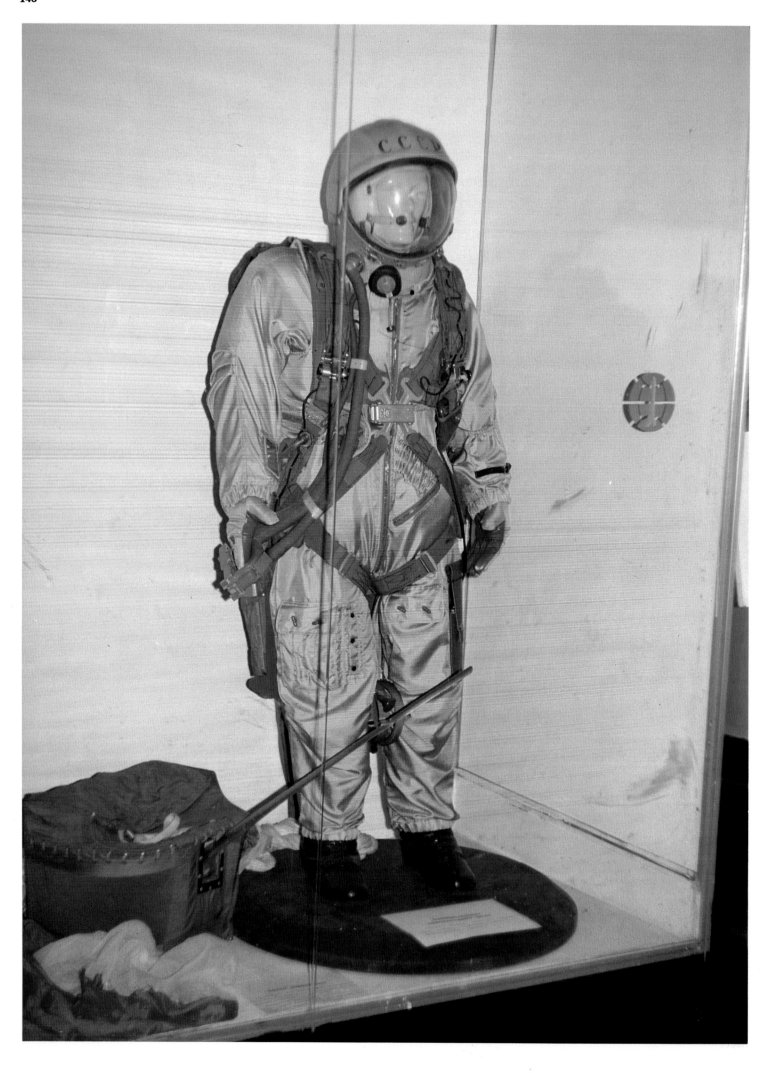

systems had advanced to a point where crews no longer wore spacesuits in the capsule. This practice continued until the Soyuz 11 disaster.

Voskhod 1 carried the first multi-man crew into space. Their brief mission concentrated on biomedical observations and extensive system checks. Live TV pictures of the crew were beamed back to Earth.

Kosmos 57

Launch Vehicle: A-2
Launch Site: Tyuratam
Launch Date: 22 February 1965
Total Weight: 5682 kilograms
Apogee: 512 kilometers
Perigee: 175 kilometers
Inclination: 64.8 degrees
Period: 91.1 minutes

The unmanned precursor for Voskhod 2, Kosmos 57 exploded in orbit before it could be recovered. The Soviets delayed their next manned launch for almost a month in the aftermath of the failure.

Voskhod 2

Launch Vehicle: A-2
Launch Site: Tyuratam
Launch Date: 18 March 1965
Recovery Date: 19 March 1965
Total Weight: 5682 kilograms
Apogee: 495 kilometers
Perigee: 173 kilometers
Inclination: 65 degrees
Period: 90.9 minutes
Crew: P Belyayev and A Leonov

Voskhod 2 contained further modifications, including an expandable airlock attached to the reentry module. The appendage weighed 250 kilograms, had a length of 2.5 meters and a diameter of 1.2 meters. The airlock allowed Lieutenant Colonel Leonov to don a spacesuit and exit the capsule without causing a loss of air in the descent module. Leonov was in the open airlock for about ten minutes, and he dangled from the end of a 5.35-meter tether outside the vehicle for another ten minutes before terminating history's first space walk. The EVA was recorded by a TV camera mounted on the reserve retrorocket assembly and a film camera mounted on the airlock. Leonov retrieved the film camera when he reentered the descent module. The suit was very cumbersome, and Leonov experienced a disorienting euphoria while outside of the vehicle. Both these factors combined to prevent him from conducting any experiments or us-

Left: Alexei Leonov's spacesuit for his historical EVA. The suit is orange for better visibility by recovery crews. Above: a Vostok cosmonaut's ejection seat.

ing a film camera which was attached to his thigh to photograph the spacecraft.

The automatic reentry system on Voskhod 2 failed and Colonel Belyayev was forced to make a manual descent. Because the reentry burn was delayed, the capsule missed the scheduled landing zone by some hundreds of kilometers, creating a delay of 24 hours before ground crew support teams could reach the crew.

Kosmos 110

Launch Vehicle: A-2
Launch Site: Tyuratam
Launch Date: 22 February 1966
Recovery Date: 16 March 1966
Total Weight: 5700 kilograms
Apogee: 904 kilometers
Perigee: 187 kilometers
Inclination: 51.9 degrees
Period: 96.3 minutes

The final Voskhod mission carried the dogs Veterok and Ugolek. TV pictures and biomedical telemetry were returned during the record 22-day mission. Data from this flight was used to plan for the longer manned missions of the Soyuz program, which was initiated the next year.

Vostok

Along with the Sputnik launches, the Vostok, or East, program epitomizes for the USSR the glory days of the early space race. By early 1960 the Soviets had accumulated enough experience with recoverable payloads and biosats containing dogs that they felt confident in placing human payloads into orbit. The Vostok vehicle consisted of two modules, a reentry capsule and an instrument module. The reentry capsule was a 2.5 meter sphere that was

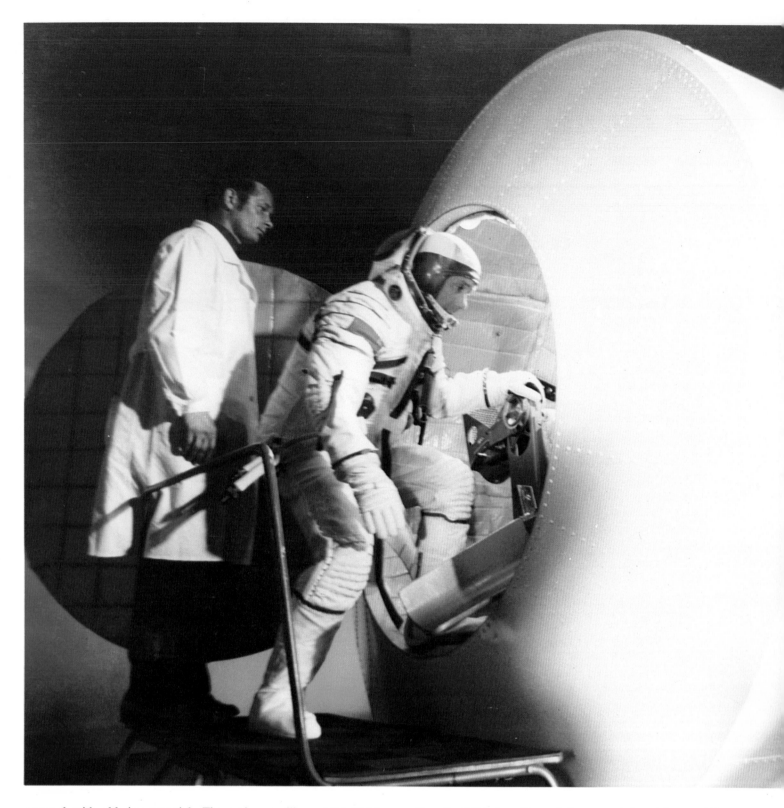

covered with ablative material. The main interior feature was the cosmonaut's ejection seat. There were three small viewing ports in the capsule, TV and film cameras, a radio system, control panels, life support equipment and food and water. Two communications antennas were mounted on top of the Vostok descent capsule. From the very beginning of their manned space program, the Soviets relied upon automatic ground control for all mission-critical activities. Manual control by cosmonauts is used only in emergencies.

The instrument module was conically shaped. It contained the instruments for con-

trolling orbital flight, as well as the TDU-1 retrorocket system for deorbiting the capsule. The TDU-1 used an amine/nitrous oxide liquid fuel engine that created 1614 kilograms of thrust. The waist of the Vostok vehicle was encircled with spherical tanks of compressed gas. The instrument module was partially covered with thermal radiation louvers.

Vostok 1

Launch Vehicle: A-1
Launch Site: Tyuratam
Launch Date: 12 April 1961
Recovery Date: 12 April 1961

Above: **A Cosmonaut steps into a centrifuge at the Gagarin Cosmonaut Training Center. The centrifuge tests reactions under stress.**

Total Weight: 4725 kilograms
Apogee: 327 kilometers
Perigee: 181 kilometers
Inclination: 65 degrees
Period: 89.1 minutes
Crew: Y Gagarin

Yuri Gagarin's historic flight began with lift-off at Tyuratam, followed by separation of the four strap-on stages of the A-1 booster 119 seconds later. The protective shroud covering Vostok 1 was jettisoned at the 156-

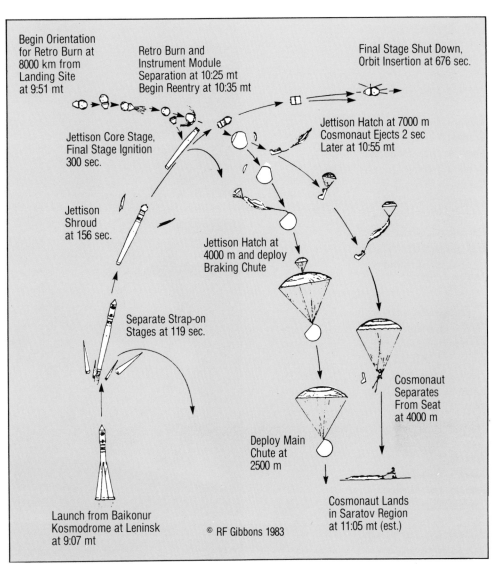

Begin Orientation for Retro Burn at 8000 km from Landing Site at 9:51 mt

Retro Burn and Instrument Module Separation at 10:25 mt Begin Reentry at 10:35 mt

Final Stage Shut Down, Orbit Insertion at 676 sec.

Jettison Core Stage, Final Stage Ignition 300 sec.

Jettison Hatch at 7000 m Cosmonaut Ejects 2 sec Later at 10:55 mt

Jettison Shroud at 156 sec.

Jettison Hatch at 4000 m and deploy Braking Chute

Separate Strap-on Stages at 119 sec.

Cosmonaut Separates From Seat at 4000 m

Deploy Main Chute at 2500 m

Launch from Baikonur Kosmodrome at Leninsk at 9:07 mt

© RF Gibbons 1983

Cosmonaut Lands in Saratov Region at 11:05 mt (est.)

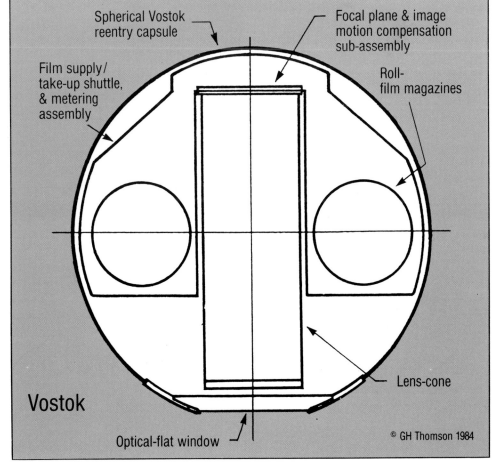

Spherical Vostok reentry capsule

Focal plane & image motion compensation sub-assembly

Film supply/ take-up shuttle, & metering assembly

Roll-film magazines

Lens-cone

Vostok

Optical-flat window

© GH Thomson 1984

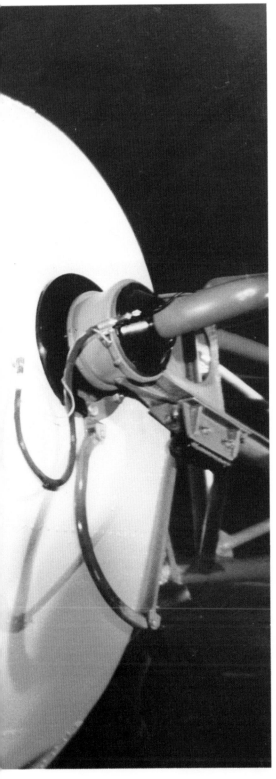

second mark. The core stage of the booster burned out around 300 seconds into the launch, and the final stage was ignited. Orbital insertion and final stage shut-down occurred simultaneously at 676 seconds into the flight. Gagarin made one orbit. The vehicle was oriented for retrorocket burn some 44 minutes into the flight. The engine was fired and the instrument module separated from the descent capsule one hour and 18 minutes into the mission, with reentry commencing ten minutes later. Following reentry, Gagarin ejected from the capsule at an altitude of 7000 meters one hour and 48 minutes after lift-off. Gagarin separated from his ejection seat at 4000 meters and descended via parachute to his landing zone in the Saratov region. The empty capsule also descended via parachute. The entire mission lasted 118 minutes.

Vostok 2
Launch Vehicle: A-1
Launch Site: Tyuratam
Launch Date: 6 August 1961
Recovery Date: 7 August 1961
Total Weight: 4731 kilograms
Apogee: 257 kilometers
Perigee: 178 kilometers
Inclination: 64.9 degrees
Period: 88.6 minutes
Crew: G Titov

The second man to reach orbit, German Titov, remained there for 24 hours—eventually completing 17 orbits before recovery. All of the Vostok cosmonauts ejected from the capsule and parachuted to Earth like Gagarin, because the capsule landed with too much residual force for the safety of human cargo.

Vostok 3
Launch Vehicle: A-1
Launch Site: Tyuratam
Launch Date: 11 August 1962
Recovery Date: 15 August 1962
Total Weight: 4722 kilograms
Apogee: 251 kilometers
Perigee: 183 kilometers
Inclination: 65 degrees
Period: 88.5 minutes
Crew: A Nikolayev

The Soviets increased their time in orbit fourfold with Vostok 3. As a precaution against malfunction, all Vostok flights were placed in orbits from which they would decay naturally in 10 days. The capsules were always stocked with 10 days supply of food and water, and life support systems were designed to function for that length of time as well. The ejection seat also contained emergency rations, as well as a dingy.

Vostok 4
Launch Vehicle: A-1
Launch Site: Tyuratam
Launch Date: 12 August 1962
Recovery Date: 15 August 1962
Total Weight: 4728 kilograms
Apogee: 254 kilometers
Perigee: 180 kilometers
Inclination: 65 degrees
Period: 88.5 minutes
Crew: P Popovich

Vostok 4 followed its predecessor off the pad within 24 hours and subsequently

approached to within 6.5 kilometers of Vostok 3. Such maneuvers demonstrated impressive capabilities with respect to mission control and pad turn-around time.

Vostok 5
Launch Vehicle: A-1
Launch Site: Tyuratam
Launch Date: 14 June 1963
Recovery Date: 19 June 1963
Total Weight: 4720 kilograms
Apogee: 222 kilometers
Perigee: 175 kilometers
Inclination: 65 degrees

Above: The Vostok craft at assembly stage. *Note* the oxygen and nitrogen life-support system bottle 'necklace.' *Right:* Vostok on display. *Note* the Kosmos-designated rocket.

Period: 88.3 minutes
Crew: V Bykovskiy

Lieutenant Colonel Bykovskiy spent 119 hours and six minutes in orbit, a duration record that was not exceeded by a Soviet spacecraft until the Soyuz 9 mission. Vostok 5 passed within five kilometers of Vostok 6, which was launched two days after Bykovskiy lifted off.

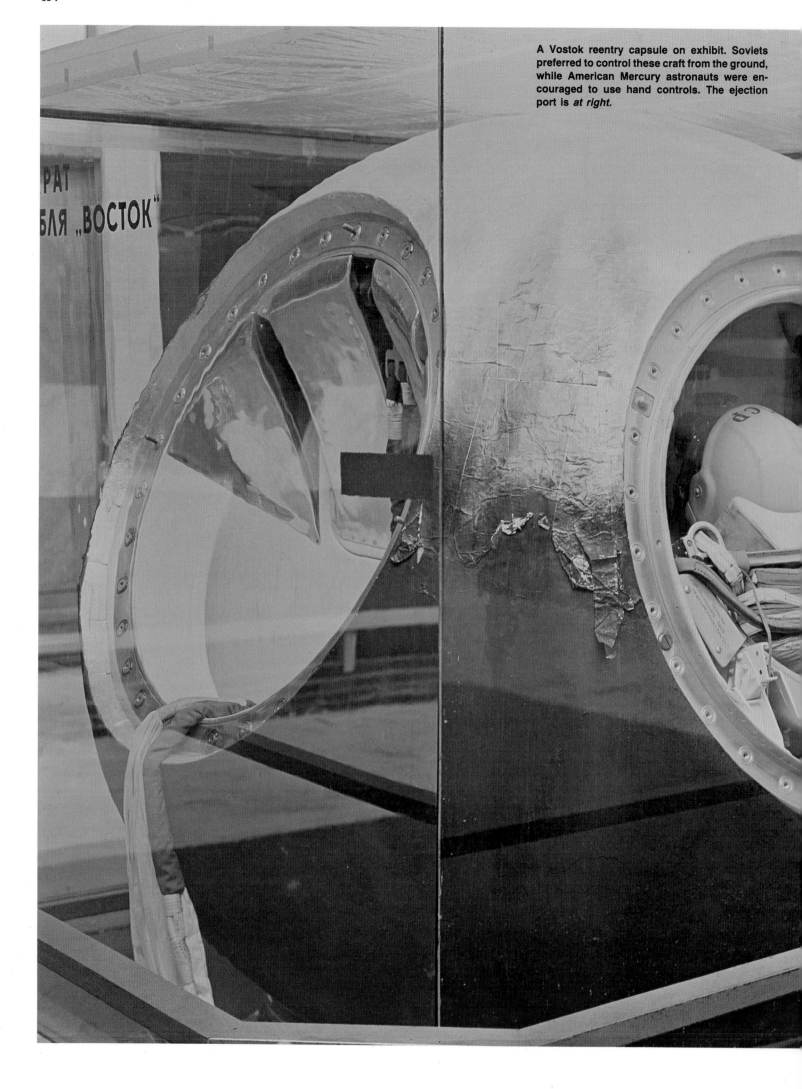

РАТ

БЛЯ „ВОСТОК"

A Vostok reentry capsule on exhibit. Soviets preferred to control these craft from the ground, while American Mercury astronauts were encouraged to use hand controls. The ejection port is *at right*.

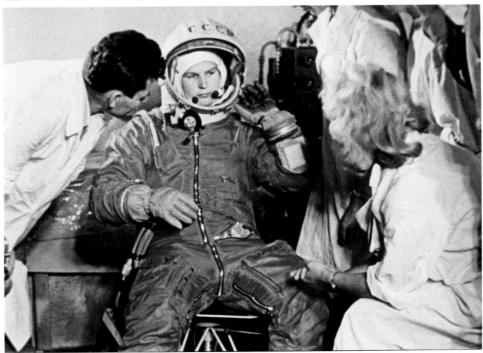

Above, foreground: Yuri Gagarin shares a joke with fellow cosmonaut Alexei Leonov.

Below: Valentina Tereshkova, the first spacewoman, before her Vostok 6 flight.

Above: The A-1 launch vehicle was developed from a Soviet ICBM. *Right:* This is the business end of a Vostock final stage. The flight module is visible *to the left.*

Vostok 6

Launch Vehicle: A-1
Launch Site: Tyuratam
Launch Date: 16 June 1963
Recovery Date: 19 June 1963
Total Weight: 4713 kilometers
Apogee: 233 kilometers
Perigee: 183 kilometers
Inclination: 65 degrees
Period: 88.3 minutes
Crew: V Tereshkova

The final Vostok flight carried the first woman into space. By the end of her mission, Valentina Tereshkova had accumulated more time in orbit than the entire US Mercury program of six manned flights. Tereshkova had two key qualities that were attractive to the Soviet authorities: She was an accomplished sports parachutist, and, as a former textile factory worker, she came from a genuine proletarian background. While her mission was certainly a spectacular success, Soviet mission planners may have had to delay the launch for a day, eliminating the chances for any joint flight with Vostok 5 other than the brief fly-by described above.

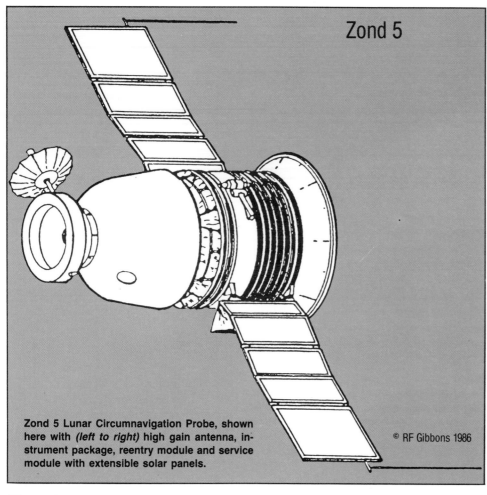

Zond 5

Zond 5 Lunar Circumnavigation Probe, shown here with *(left to right)* high gain antenna, instrument package, reentry module and service module with extensible solar panels.

© RF Gibbons 1986

Zond

The Zond program did not have a single central focus, since Zond launches sent spacecraft on lunar, interplanetary and near-Earth orbit missions. Three distinct phases of Zond launches occurred between 1964 and 1970: a Venus mission, two Mars missions and five manned lunar precursor missions. These space flights, and their related Kosmos failures, are described in this section.

Zond 1

Launch Vehicle: A-2e
Launch Site: Tyuratam
Launch Date: 2 April 1964
Fly-By Date: 19 July 1964
Total Weight: 890 kilograms
Apogee: 213 kilometers
Perigee: 187 kilometers
Inclination: 64.8 degrees
Period: 88.5 minutes
Aphelion: 1.001 astronomical units
Perihelion: 0.652 astronomical units
Inclination: 3.7 degrees
Period: 274 days

Zond 1 was a partially successful Venus mission that came between Venera 1 and Venera 2 (see the Venera entry for details). Although the spacecraft passed within 100,000 kilometers of Venus in mid-July

1964, it ceased transmitting to ground stations on 14 May.

Zond 2

Launch Vehicle: A-2e
Launch Site: Tyuratam
Launch Date: 30 November 1964
Fly-By Date: 6 August 1965
Total Weight: 890 kilograms
Apogee: 219 kilometers
Perigee: 153 kilometers
Inclination: 64.7 degrees
Period: 88.2 minutes
Aphelion: 1.52 astronomical units
Perihelion: 0.98 astronomical units
Inclination: 6.4 degrees
Period: 508 days

The second and third Zond missions were sandwiched between Mars 1 and 2 (see the Mars entry for details). Zond 2 suffered a communications failure common to early Soviet interplanetary probes in April 1965. Its closest approach to Mars, in early August, was 1500 kilometers.

Zond 3

Launch Vehicle: A-2e
Launch Site: Tyuratam
Launch Date: 18 July 1965
Total Weight: 960 kilograms
Apogee: 210 kilometers
Perigee: 164 kilometers

Inclination: 64.8 degrees
Period: 88.2 days
Aphelion: 1.56 astronomical units
Perihelion: 0.9 astronomical units
Inclination: 0.5 degrees
Period: 500 days

Zond 3 was launched outside of the Mars launch window as a diagnostic test for interplanetary probe's communications systems. On the way to intersecting with the Martian orbital plane, the vehicle flew by the moon at a distance of 9200 kilometers and collected 25 photographs of the far side. These photographs were then sent back to Earth via facsimile representation at increasing distances as a means of testing the vehicle's communications capability.

Zond 4

Launch Vehicle: D-1e
Launch Site: Tyuratam
Launch Date: 2 March 1968
Total Weight: 3375 kilograms
Apogee: 206 kilometers
Perigee: 190 kilometers
Inclination: 51.5 degrees
Period: 88.4 minutes
Apcynthion: 400,000 kilometers

Intended as an engineering and diagnostic test for future Zond flights, this mission was launched outside the optimal lunar launch window in the opposite direction from the Moon, so that its return would be unaffected by lunar gravity. The craft itself was essentially identical to the early Soyuz variants, with the exception that the forward module had been removed. The vehicle itself weighed some 4280 kilograms, had a length of 5.3 meters and a width of 2.3 meters. Zond 4 constituted the first use of the D-1e booster in the program.

Zond 5

Launch Vehicle: D-1e
Launch Site: Tyuratam
Launch Date: 14 September 1968
Recovery Date: 21 September 1968
Total Weight: 5375 kilograms
Apogee: 219 kilometers
Perigee: 187 kilometers
Inclination: 51.3 degrees
Period: 88.4 degrees
Apcynthion: 400,000 kilometers

This mission was sent around the moon, passing it at a distance of 1950 kilometers for return and recovery at night in the Indian Ocean. The Soviets used the occasion of their first deep space recovery to accomplish several important objectives. Zond 5 was the first Soviet water landing and constituted a very useful precursor for any manned lunar missions which had to land outside the Soviet Union. The vehicle also returned high quality imagery of the far side the moon. Finally, Zond 5 was a biosat, containing turtles, meal worms, wine flies, lysogenic bacteria, chlorella, a spiderwort plant, as well as wheat, pine and barley seeds. The biological payload was contained in a cabin component that also held scientific instruments, communications equipment, power supply apparatus, heat regulation devices and the parachutes for descent following reentry. A second module held the flight control instrumentation, solar panels, telemetry system, batteries and more heat regulation devices.

Zond 6

Launch Vehicle: D-1e
Launch Site: Tyuratam
Launch Date: 10 November 1968
Recovery Date: 17 November 1968
Total Weight: 5375 kilograms
Apogee: 210 kilometers
Perigee: 185 kilometers
Inclination: 51.4 degrees
Period: 87.9 minutes
Apcynthion: 400,000 kilometers

Zond 6 rounded the Moon at 2250 kilometers and returned to Earth using a skip-reentry profile in order to land inside the Soviet Union. The spacecraft carried a biological payload, micrometeorite detection equipment and instruments for studying the paths of cosmic rays. Cameras on board Zond 6 returned high quality stereo imagery of the lunar surface. After the successful completion of this mission the Soviets announced that the last three Zond flights had been precursors for manned circumlunar efforts.

Zond 7

Launch Vehicle: D-1e
Launch Site: Tyuratam
Launch Date: 7 August 1969
Recovery Date: 14 August 1969
Total Weight: 5375 kilograms
Apogee: 191 kilometers
Perigee: 183 kilometers
Inclination: 51.5 degrees
Period: 88.2 minutes
Apcynthion: 400,000 kilometers

This mission was essentially identical to that of Zond 6, with the exception that color imagery was obtained of the Earth and the Moon, as well as black-and-white photographs.

Below: A mockup of Zond 3, which, in adding to the photos taken by Luna 3, nearly completed the mapping of the Moon's far side while en route to entering a solar orbit.

Zond 8

Launch Vehicle: D-1e
Launch Site: Tyuratam
Launch Date: 20 October 1970
Recovery Date: 27 October 1970
Total Weight: 5375 kilograms
Apogee: 223 kilometers
Perigee: 202 kilometers
Inclination: 51.5 degrees
Period: 88.7 minutes
Apcynthion: 400,000 kilometers

The last Zond mission circled the Moon at a distance of 1100 kilometers, while collecting both color and black-and-white imagery of the lunar surface. The spacecraft also broadcast the first TV pictures of the Earth from 65,000 kilometers. The major innovation during this mission, however, was the craft's reentry profile, which used a North Pole trajectory (all other Zond recoveries had commenced with reentry over the South Pole) without the usual skip-return profile, and culminated in a nighttime-splash-down in the Indian Ocean. Although the Soviets clearly gained useful experience from this mission and its predecessors, they did not take the next step toward a manned lunar landing by placing a human crew in lunar orbit, probably due to difficulties experienced in developing heavy boosters (see Appendix II for details).

APPENDIX I:
SOVIET SPACE STATIONS

Two areas of indisputable Soviet leadership in space are manhours in orbit and space biomedical applications. Soviet strength in these two areas comes directly from their emphasis, since the early 1970s, upon low-Earth orbit laboratories for their manned space program. Since 1971 the USSR has placed three generations of space stations into orbit and has developed a series of manned transport and unmanned resupply vehicles to service them. The Soyuz and Progress craft that bring men and supplies to Soviet space stations are described in the main body of this encyclopedia, along with synopses of the activities of each crew to occupy the orbital complexes through the end of 1985. This appendix describes both the Salyut- and Mir-type space stations, summarizes the experiments that have been conducted aboard them and examines their ground control support systems.

Salyut

The Salyut, or Salute, series of Soviet spacecraft can be divided into two generations. Salyut 1-5 were labeled the first generation by Western experts, and Salyut 6 and 7 the second. The distinguishing external feature between the two generations was one docking port for the first generation stations and two for the second. This section of the appendix describes the Salyut craft in general and also provides a synopsis of key events associated with the operational lives of each of the seven Salyut space stations.

The Salyut main body is 13 meters in length, and its maximum diameter is 4.2 meters. The vehicle weighs 19 metric tons. Salyut is made up of three major components: a docking and airlock component, the laboratory/living quarters and the control instrumentation and propulsion module. The first two areas are pressurized and, in combination, offer some 100 cubic meters of habitable space for up to five crew members. The Salyut station, however, should be viewed as one module in a complex usually composed of itself, a Soyuz ferry craft and a Progress or Progress follow-on resupply vehicle/space tug. The total length of such a

configuration is 29 meters. The space station is powered by a series of solar panels deployed at right angles from the main body of the vehicle.

The docking and airlock component (the forward docking module for the second generation Salyuts) contains the passive docking collar for Soyuz craft, an access hatch for ground servicing and EVA, seven portholes and two control posts for station orientation. Inside the airlock are spacesuits and other equipment used during EVAs. The following apparatus are attached to the outside of the docking module:

· docking-related lamps, antenna and TV cameras
· temperature regulation panels
· compressed air bottles
· solar and ionic orientation sensors
· EVA handrails
· panels for studying micrometeorites and the effects of vacuum upon rubber, optical surfaces and biological polymers

Moving aft from the airlock, the living quarters/operations area is divided into a series of control posts. Post No 1 has two working positions from which the main communications, life support and other ship systems are monitored and controlled. Post No 2 is devoted to astro-navigation and space station orientation activities. Between these two main control stations is a table configured for eating in zero gravity and carrying out minor repairs. Further aft is Control

Post No 3, which controls the various scientific apparatus on board the ship, such as the materials processing devices and radio telescope. Biomedical observations and experiments are conducted at Post No 4, which is equipped with a treadmill, bicycle ergomometer, a pneumatic vacuum suit and muscle stimulation apparatus. This area also contains still, motion picture and TV cameras. The East German MKF-6M multispectral camera system is mounted at a porthole near Post No 4 (there are a total of 19 portholes in the Salyuts). Bunks, a shower device and the head are located at the aft end of the living quarters/operations module.

The major structures outside the living quarters/operations area are the solar panels, which, on the second generation stations, cover a total of 60 square meters. In addition to the solar panels, automatic orientation sensors (infrared, solar and TV) are located in this area. Finally, a device attached to the outside of the MKF-6M porthole operates a repeating-action cover in order to maintain a specific temperature range for the camera's porthole.

The next module aft is devoted to housing the bulk of the station's scientific equipment. During 14 years of Salyut operations thousands of experiments have been conducted, millions of square kilometers of the Earth's surface have been photographed and major contributions have been made to the fields of astronomy and space medicine. The Kristall, Splav and Insparitel furnaces and ovens have been used for the bulk of the

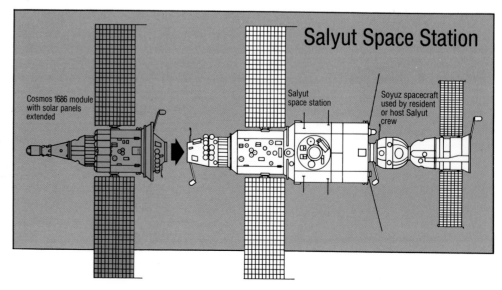

Salyut Space Station

Cosmos 1686 module with solar panels extended

Salyut space station

Soyuz spacecraft used by resident or host Salyut crew

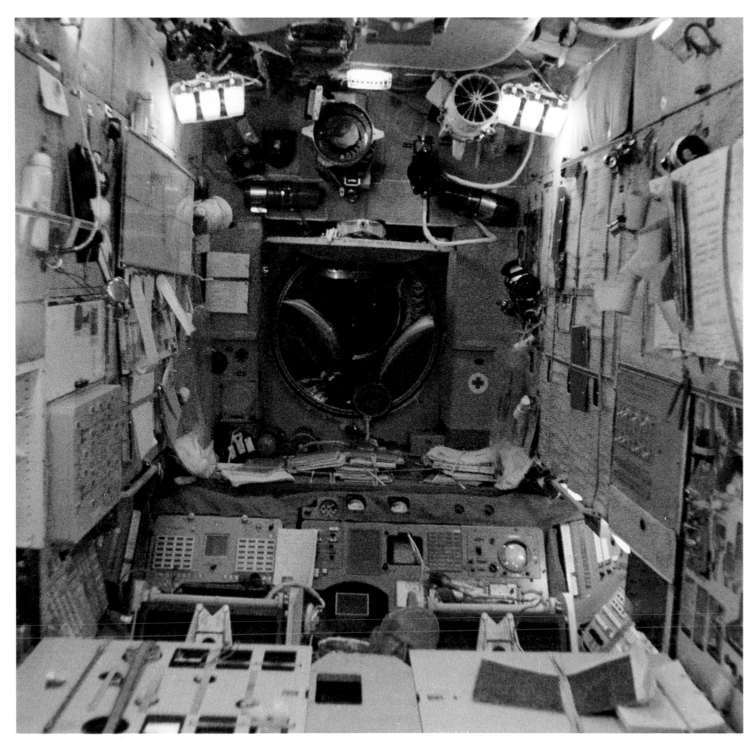

materials processing program to produce homogeneous alloys, semiconductors and vaccines. The Earth resources observations conducted from Salyut and unmanned vehicles are estimated by the Soviets to save their economy billions of roubles per year. Radio and infrared telescopes have been used to study the Sun and other stars, as well as the planets, from outside the distorting effects of the Earth's atmosphere. The lengthy periods in space spent by the long-duration crews aboard the Salyuts have supplied the Soviets with unrivaled data concerning the effects of zero-gravity upon the human anatomy.

The aft end of the Salyut contains the attitude control thrusters and the main engine, which is capable of generating some 300 kilograms of thrust. Supplies delivered

by Progress and Progress follow-on vehicles are stored in this area. On the second generation Salyuts, this section also contains a second docking port, from which the station can be refuelled. The equipment deployed outside of this module is similar to the docking support apparatus described for the forward port.

The major systems associated with the first and second generation Salyuts are described below:

Onboard equipment control system (Soviet acronym: SUBK): This system operates in

three modes: automatic, radio control from Earth and via command from the crew. The SUBK controls the station's power supply, external element deployment sequences and data links to ground control.

Orientation and motion control system (SOUD): This system is responsible for maintaining the orientation of the Salyut complex in either an automatic or manual mode. It integrates data from all the various sensors deployed on the Salyut stations, as well as controlling the docking lamps and targets.

Combined motor installation (ODU): This system controls the low thrust orientation and the high-thrust maneuvering engines.

Command radio link and television communications system: This system handles command uplink from ground control to the stations, trajectory information, telemetry and telecommunications between the stations and ground control.

Telephone communications systems (Zarya): Zarya differs from the system described above in that it handles short wave and ultrashort wave communications between ground control and the space stations and between the stations and their Soyuz ferry craft. Zarya also handles the transmission of alphanumeric text between stations and ground control.

Radiotelemetry communications systems (RTS): The RTS collects and downlinks telemetry associated with space station operations and scientific experiments. Back-up magnetic tapes for the experiments are returned on Soyuz ferry craft or in Progress follow-on reentry capsules.

Power supply system (SEP): The SEP supplies power to the station module and any craft docked with the Salyut. The system includes those elements that maintain the proper orientation for the solar panels toward the Sun, and those which charge and maintain the buffer batteries that supply emergency power in case of a panel failure.

Life support system (SOZh): This system controls internal atmosphere content and pressure, food and water supplies, hygiene systems and EVA support. Critical subsystems include gas composition and support for regenerating oxygen and absorbing carbon dioxide (SOGS); water regeneration via atmospheric moisture condensation; and a sanitary-hygienic installation which includes a shower and head (ASU).

Medical monitoring and prophylaxis equipment: This system includes examination and diagnostic equipment that return biomedical data on the cosmonauts to Earth-side medical facilities. It also contains exercise equipment used by the crew to maintain body tone while in orbit and to prepare for readjustment to Earth's gravity upon return.

Temperature regulation system (STR): The STR both maintains temperatures for the living quarters of the Salyuts and monitors and maintains temperatures of the scientific apparatus and the transport vehicles docked to the stations in a powered-down condition.

Right: **Technicians converse through the open EVA hatch of this Salyut display model at the Gagarin Cosmonaut Training Center.** *Note* **the Soyuz docking probe positioned** *at right.* *Below:* **Vladimir Kovalyonok** *(left)* **and Alexander Ivanchenkov are shown here aboard the Soyuz 29-Salyut 6 space complex.**

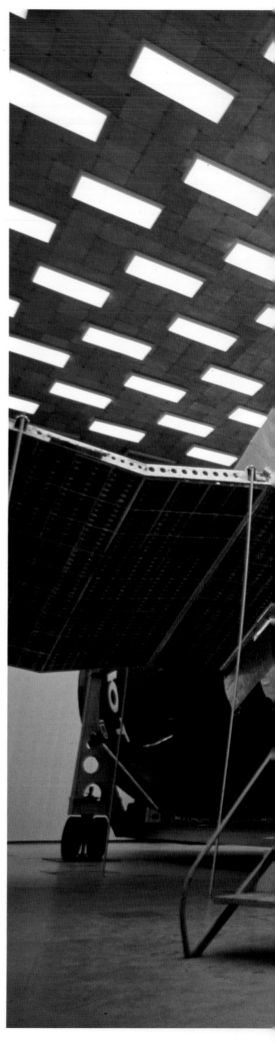

Docking and internal transfer system (SSVP): This system is the space station's component of the automatic docking system. It performs mechanical, electrical and hydraulic docking functions. In addition, the SSVP controls the airlock environment.

The seven Salyut space stations and the related launch of Kosmos 557 are described below:

Salyut 1

Launch Vehicle: D-1
Launch Site: Tyuratam
Launch Date: 19 April 1971
Decay Date: 11 October 1971
Total Weight: 18,300 kilograms
Apogee: 222 kilometers
Perigee: 200 kilometers
Inclination: 51.6 degrees
Period: 89.6 minutes

The first Salyut was used as an observation platform in order to gather data in the fields of astronomy, earth resources and meteorology. Two types of telescope were carried aboard Salyut 1. One of these instruments, called Orion 1, was designed to collect spectrograms of stars in the region of 2000 to 3000 angstroms in the electromagnetic spectrum. The other was a gamma-ray telescope, dubbed Anna III by the Soviets, which was capable of making observations of the cosmos that are impossible from the Earth's surface. Earth resources and meteorological photography was coordinated with the product of other spaceborne and air-breathing collection systems. A hydroponic program studied the effects of microgravity on growing plants, and numerous biomedical observations were made of the cosmonauts themselves, in order to determine the effects of prolonged weightlessness on human physiology, proving their expertise in space medicine.

Salyut 2

Launch Vehicle: D-1
Launch Site: Tyuratam
Launch Date: 3 April 1973
Decay Date: 28 May 1973
Total Weight: 18,600 kilograms
Apogee: 260 kilometers
Perigee: 215 kilometers
Inclination: 51.6 degrees
Period: 89 minutes

Salyut 2 was probably the first of the so-called military Salyuts. Western experts base this judgement upon the vehicle's use of telemetry similar to that of Soviet photo-reconnaissance satellites operating at the time. The exact nature of the second Salyut's mission will probably never be known in the West, since the vehicle suffered a catastrophic explosion on 14 April which tore the solar panels, docking apparatus and the radio transponder from the craft and left it tumbling helplessly in a low-Earth orbit from which it decayed in late May 1973.

Kosmos 557

Launch Vehicle: D-1
Launch Site: Tyuratam
Launch Date: 11 May 1973
Decay Date: 22 May 1973
Total Weight: 18,600 kilograms
Apogee: 226 kilometers
Perigee: 218 kilometers
Inclination: 51.6 degrees
Period: 89.1 minutes

The next Salyut-type launch appears to have been a failure of a civilian space station. Kosmos 557 used the manned program

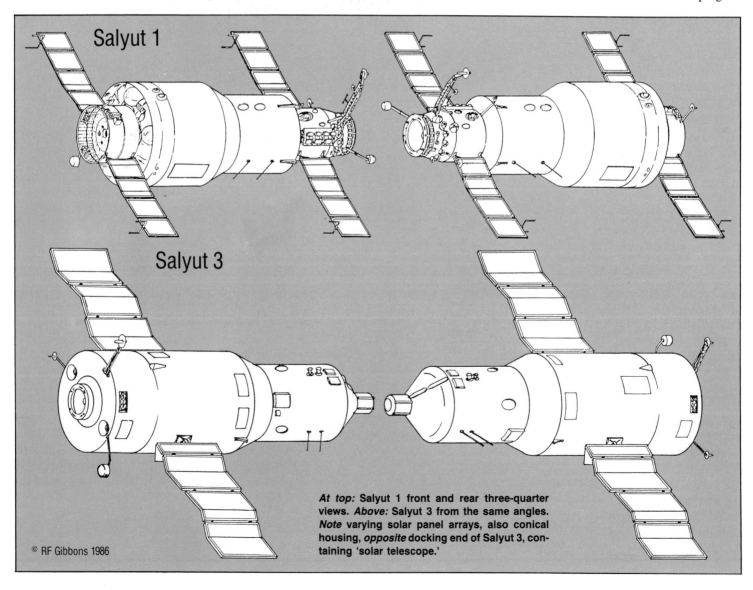

© RF Gibbons 1986

At top: Salyut 1 front and rear three-quarter views. **Above:** Salyut 3 from the same angles. **Note** varying solar panel arrays, also conical housing, **opposite** docking end of Salyut 3, containing 'solar telescope.'

telemetry channels associated with Salyut 1, but the vehicle apparently failed so early in its flight—and did so during the obvious failure of Salyut 2—that the Soviets gave the craft the usual Kosmos designation for an obituary.

Salyut 3

Launch Vehicle: D-1
Launch Site: Tyuratam
Launch Date: 24 June 1974
Decay Date: 24 August 1975
Total Weight: 18,600 kilograms
Apogee: 270 kilometers
Perigee: 219 kilometers
Inclination: 51.6 degrees
Period: 89.1 minutes

The first operational military Salyut, this craft had three major design modifications vis-a-vis Salyut 1. The fore- and aft-fixed solar panels were replaced by two large steerable panels mounted just aft of the middle of the vehicle. The docking port was moved to the rear of the station. Finally, in keeping with the photo-reconnaissance mission of Salyut 3, the station was equipped with a reentry capsule for film that was

attached to the forward end of the craft. Salyut 3 used military reconnaissance telemetry, and very few photographs were ever released detailing activities aboard the station or the areas of the Earth which were imaged. In addition to their military photography, the crew also conducted some 400 scientific experiments.

Salyut 4

Launch Vehicle: D-1
Launch Site: Tyuratam
Launch Date: 26 December 1974
Decay Date: 2 February 1977
Total Weight: 18,600 kilograms
Apogee: 270 kilometers
Perigee: 219 kilometers
Inclination: 51.6 degrees
Period: 89.1 minutes

Salyut 4 continued the alternating series of civilian and military missions. This station contained a great deal more scientific equipment than its predecessors but continued the

tripartite research program involving astronomy, Earth resources and biomedical observations. Three types of spectrometric telescopes (infrared, solar and x-ray) were used for observations of the solar system, the Milky Way and other galaxies. Earth resources photography was devoted to enhancing the output of the agricultural, timber and fishing sectors of the Soviet economy. Micro-organisms and plants were grown aboard the space station. Biomedical investigations concentrated upon the effects of microgravity on the human cardiovascular system.

Salyut 5

Launch Vehicle: D-1
Launch Site: Tyuratam
Launch Date: 22 June 1976
Decay Date: 8 August 1977
Total Weight: 18,600 kilograms
Apogee: 260 kilometers
Perigee: 219 kilometers
Inclination: 51.6 degrees
Period: 89 minutes

Salyut 5 was the last of the dedicated military space stations. Like Salyut 3, it employed military telemetry channels and jetti-

Below: left to right: The crew of Soyuz 38, Cuban cosmonaut Arnoldo Mendez and Soviet cosmonaut Yuri Romanenko, aboard Salyut 6.

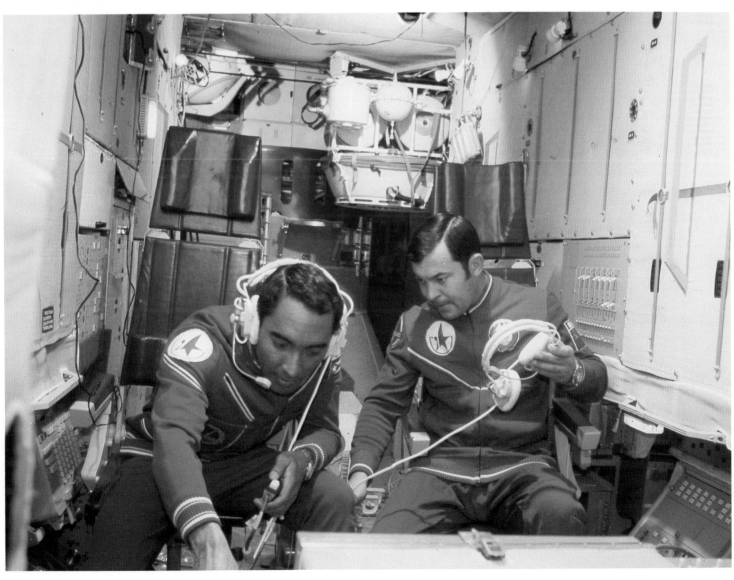

The uses of the cylinder: This Salyut 7 photo is a study in perspective. *Note* the rail dollies under the craft, and the D-1 launch vehicle with strap-on boosters, also on rails, *at left*.

Left: **Svetlana Savitskaya discusses equipment with manned-spacecraft technicians aboard a training mockup of Salyut 7.**

soned a reentry capsule following the departure of its final crew. In addition to conducting military reconnaissance activities, this was the first Salyut station to employ the Splav and Kristall materials processing devices. Bismuth, tin, lead and cadmium were smelted in Splav, and a crystal growth experiment was carried out in Kristall in order to determine the prospects for manufacturing in a zero-gravity environment. Biological experiments were conducted using fish, plants, fruit flies and algae.

Salyut 6
Launch Vehicle: D-1
Launch Site: Tyuratam
Launch Date: 29 September 1977
Decay Date: 29 July 1982
Total Weight: 18,900 kilograms
Apogee: 275 kilometers

Perigee: 219 kilometers
Inclination: 51.6 degrees
Period: 89.1 minutes

Salyut 6 inaugurated the second generation of Soviet space stations. Many of the devices and procedures tested on the first five Salyuts were made operational for this mission. The station lasted more than double its originally designed operational lifetime of 18 months—primarily because the Soviets were able to refuel and resupply the vehicle and because the crews were able to conduct extensive repairs both inside and outside the vehicle. The crews devoted a third of their research time to materials processing, a third to Earth resources observations and split the last third between astrophysical and biomedical observations. The materials processing furnaces produced infrared-sensitive semiconductors, superconductors, alloys, oxides, eutectics, glass, pure metals and ionic crystals. A new multi-spectral camera was em-

ployed for Earth resources photography. A radio telescope was deployed out of the rear docking port (second generation Soviet space stations were equipped with forward and aft docking ports, facilitating resupply while occupied) in order to map the Milky Way galaxy. Design innovations included a new propulsion system, a water regeneration system and new EVA spacesuits and equipment.

Salyut 7
Launch Vehicle: D-1
Launch Site: Tyuratam
Launch Date: 19 April 1982
Total Weight: 18,900 kilograms
Apogee: 261 kilometers
Perigee: 213 kilometers
Inclination: 51.6
Period: 89.2 minutes

The last Salyut experienced a mixture of success and failure over the course of its

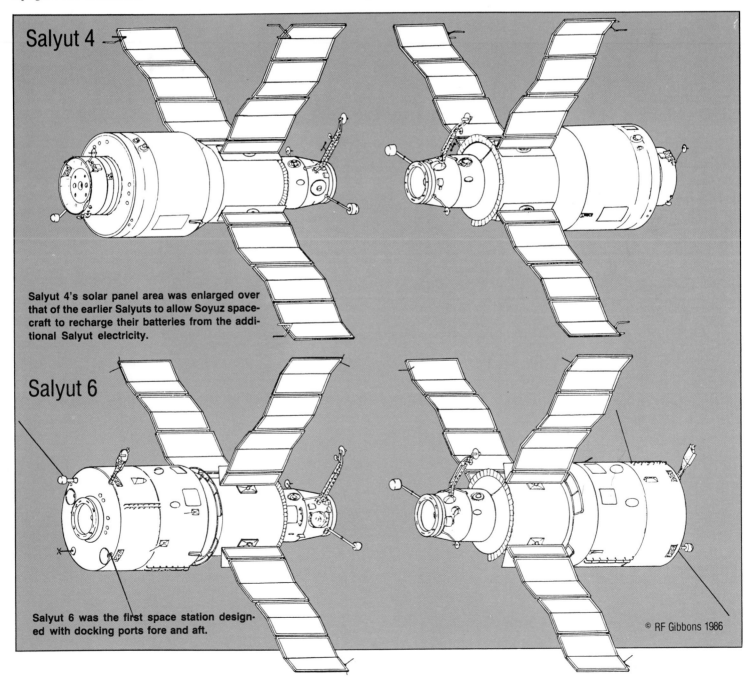

Salyut 4

Salyut 4's solar panel area was enlarged over that of the earlier Salyuts to allow Soyuz spacecraft to recharge their batteries from the additional Salyut electricity.

Salyut 6

Salyut 6 was the first space station designed with docking ports fore and aft.

© RF Gibbons 1986

operational life. Based on over four years of experience with Salyut 6, Soviet designers made various modifications to the interior of the station in order to increase crew comfort and efficiency. Portholes close to the attitude and maneuvering engines were given retractable shields in order to improve visibility when the engines are functioning. The variety of the crew's rations was expanded by the installation of a refrigerator and the color scheme for the station's interior was changed.

Improvements were also made in the amount and type of scientific equipment placed aboard Salyut 7. Materials processing was enhanced by the use of a 135 kilogram furnace, which could operate automatically when the station was unmanned. Astronomical observations were assisted by the addition of an x-ray telescope system. Biomedical apparatus were improved so that readings were data-linked to doctors on Earth in real time.

Salyut 7's catastrophic systems failure and subsequent revitalization are described in the main body of the encyclopedia under the Soyuz entry. On 6 May 1986 the space station again made history when the crew of Soyuz T-15 left the new Mir station and rendezvoused and docked with Salyut-7. The Soviets described the flight as the first use of a 'space taxicab.' Twenty days after the

Above: In the space complex 'Salyut 6, Soyuz 37 and 38', Valeri Ryumin prepares for a medical exam. *Right:* Space-gardener Svetlana Savitskaya, the first woman to fly twice in space and the first woman EVA cosmonaut, here studies the growth of plants in orbit.

reoccupation of Salyut-7, an article appeared in the Soviet journal *New Times* by Vladimir Shatalov, the head of the Soviet cosmonaut training program. He argued that this transfer operation was proof of concept for an idea originally advanced by Tsiolkovsky and Korolev in which space settlements would consist of numerous unattached vehicles capable of communication and crew exchanges. It may well be that the Soviets plan to operate a complex of several Salyut and Mir stations as an interim step towards a large integral space station. Shatalov indicated that such a complex would soon be in operation housing a total of six to 12 cosmonauts on a permanent basis. As a practical application for the near future, Shatalov described the retrieval from geosynchronous orbit of communications satellites by unmanned spacecraft, followed by repair and refurbishment at space stations in near-Earth orbit, and then return to station, thus avoiding the high cost of either replacement or retrieval and relaunch from the surface of the planet.

Soviet space station operations are controlled from a facility located at Kaliningrad,

At left: Valeri Ryumin does a wiring check aboard the Salyut 6-Soyuz 35 and 36 complex.

Above: Earth's southern hemisphere, as space photographed from Soyuz 26. *Below:* Salyut 7.

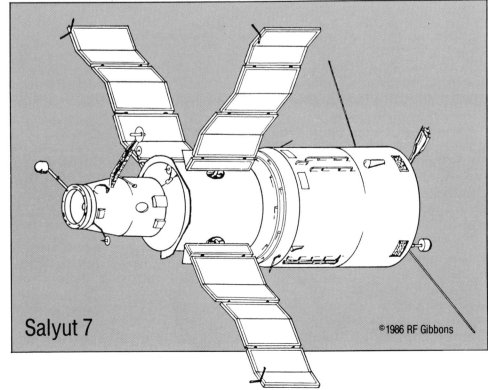

Salyut 7

©1986 RF Gibbons

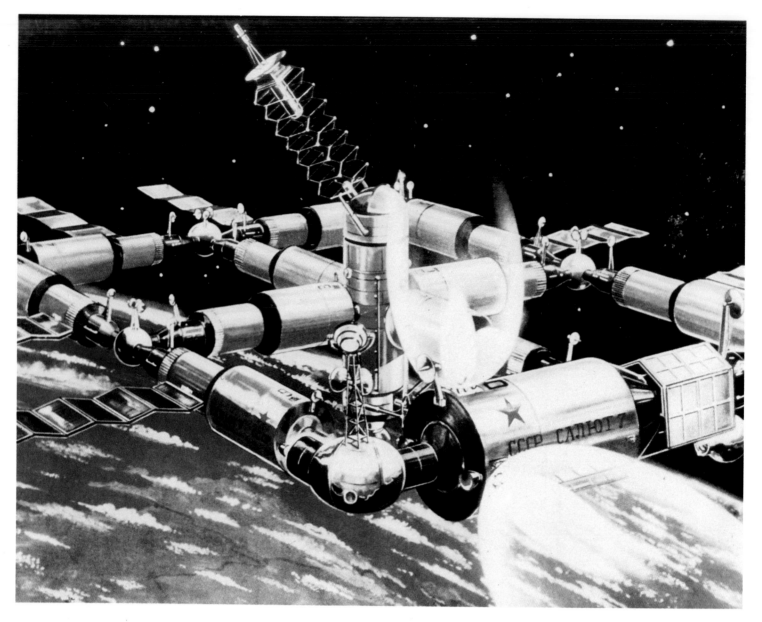

in the Moscow suburbs. Operations are supported by ground and sea-based tracking stations; land-line and Molniya satellite communications systems; and detailed physical and mathematical models of the space stations. The flight control center has two major control rooms: one controls station operations, and the other has responsibility for transport spacecraft when they rendezvous and dock with a space station. The control rooms are apparently similar to those at NASA's mission control in Houston, Texas. Work is accomplished during four shifts over a 24 hour period. During rendezvous and docking, operational control of the mission is given to the transport and cargo spacecraft control room, but once the crew enters the space station the main operations control room takes over. The following types of personnel can be found during operational shifts in the control rooms, each seated at a work position displaying data appropriate to his or her specialty.

· shift leader
· onboard systems specialists

· tracking station specialists
· flight program planners
· cosmonauts for crew liaison
· space station and spacecraft design bureau representatives
· ballisticians
· physicians
· control center systems specialists

Mir

Launch Vehicle: D-1
Launch Site: Tyuratam
Launch Date: 19 February 1986
Apogee: 359 kilometers
Perigee: 304 kilometers
Inclination: 51.6 degrees
Period: 90.9 minutes

Mir, or Peace, has been described as similar to Salyut in design, but with two solar panel arrays, instead of three, and a large docking module attached to the forward end with five ports. An additional port at the aft end of the vehicle brings the total number of

Above: An artist's conception of an advanced Soviet space station. Despite its 'pickup sticks' appearance, such a station in Earth orbit would be a powerful apparatus for both civilian and military research.

craft that can be docked with the complex to six. The Soviets may dock a module dedicated to a separate discipline (materials processing, astrophysics, Earth resources, etc) to any one of the four forward docking ports. The Soviets also plan to have international crews aboard Mir. The French recently signed an accord for a long-duration (30 day) mission involving a French spationaut, who would conduct technological and biological experiments, as well as an EVA to install a French-built cometary dust collector. Like the Salyuts, Mir, despite its name, will almost surely have some military functions. The US Defense Department has pointed out that permanently manned Soviet space stations could conduct the following missions: reconnaissance, command and control, anti-satellite operations, ballistic missile defense support operations and military satellite maintenance and control.

MIR

This illustration shows varying vehicle and module docking arrangements with the Mir space station. Its five-port docking compartment could well enable Mir to be the crucial factor in turning what was once pulp science fiction into reality. The seemingly ethereal notion of cities in orbit may well be given grounding with complex research facilities having long-term residential capabilities such as Mir. The new station will probably be used for a

continued lower right

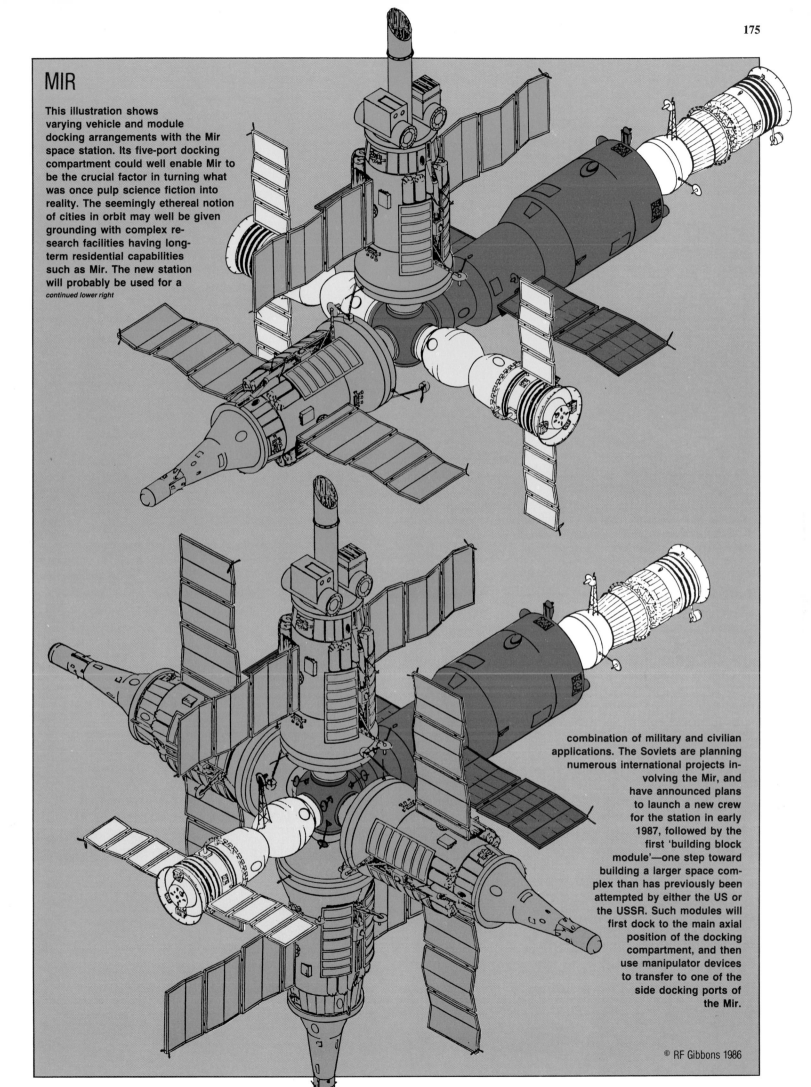

combination of military and civilian applications. The Soviets are planning numerous international projects involving the Mir, and have announced plans to launch a new crew for the station in early 1987, followed by the first 'building block module'—one step toward building a larger space complex than has previously been attempted by either the US or the USSR. Such modules will first dock to the main axial position of the docking compartment, and then use manipulator devices to transfer to one of the side docking ports of the Mir.

APPENDIX II:

SOVIET LAUNCH VEHICLES

One of the major strengths of the Soviet space program from its earliest days has been its fleet of expendable boosters. In many cases the USSR has gotten double milage out of its launch vehicles, because they were developed as variants of Soviet ICBMs. The current stable of boosters allows the USSR to maintain a much larger annual launch schedule than the United States, to compensate for vehicle failures through accelerated launch procedures and to maintain a large reserve of launch vehicles as a hedge against crisis or wartime expansion of military space activities. In their development philosophy the Soviets have followed their usual incremental approach in creating their booster fleet. Once a launch vehicle enters operational use, the basic design is generally modified several times in order to produce boosters tailored to specific missions. The most famous example of this design approach is the A-type booster, which carried the first Sputnik into orbit nearly three decades ago, but variants of which continue to launch every Soviet manned spacecraft and military photo-reconnaissance satellite.

This appendix describes each major class and variant of the Soviet launch vehicle fleet. Two basic types of alphanumeric designation are used in the West to delineate Soviet boosters. This work employs the system devised by the late Charles Sheldon of the US Congressional Research Service. This system assigns a capital letter to the basic vehicle first stage in order of historical appearance. Next, a number is given for the principal upper stage, and it is changed when a new variant, using a different upper stage, is deployed. Finally, a lower case letter is added if a final stage (eg, one that accompanies the vehicle into Earth orbit) is present. This last letter usually indicates the basic capability of the final stage (eg, e=escape, m=maneuvering, r=reentry, etc). The other nomenclature system used for Soviet launch vehicles was devised by the US Department of Defense and involves giving each booster variant an SL (for space launch) number. Thus the booster called the A-2e under the Sheldon system is referred to as the SL-6 by the US government. Table 6, along with presenting other relevant data, gives all the SL equivalents of the Sheldon system.

A-Class Launch Vehicles

A: The A, or standard launch vehicle, series has been the workhorse of Soviet boosters. Initially developed by the Korolev Design Bureau as an ICBM, the vehicle's first test took place from Tyuratam on 3 August 1957.

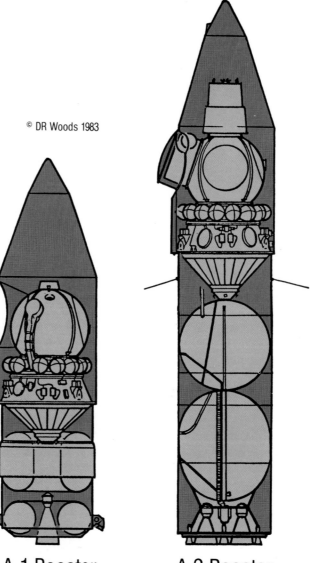

© DR Woods 1983

A-1 Booster
Vostok 4725 kg

A-2 Booster
Voskhod 5682 kg

A-2 Booster
Soyuz 6850 kg

Above: In a scene familiar to Soviet personnel, the sun's near-horizon light gilds a launch booster as it is being raised into firing position at the Tyuratam site.

Soviet Space Launch Vehicles

Vehicle	SL. No.	ICBM Antecedent	Length[1]	Diameter[1]	Total Thrust[2]	Lift Capability[3] (To 200 Km)	IOC	Payloads Launched
A	SL1-2	SS-6	31.87	10.3	504	2,000	1957	Sputnik 1–3
A-1	SL-3	SS-6	33.68	10.3	509	5,000	1959	Luna 1–3, Korabl Sputnik, Vostok, Elektron, Meteor, Kosmos (Elint, Photint, Weather)
A-2	SL-4	SS-6	42.9	10.3	534	7,500	1963	Voshkod, Kosmos (Precursors, Photint, Scientific, Resupply), Soyuz, Soyuz T, Progress
A-2e	SL-6	SS-6	42.9	10.3	540.35	7,500	1961	Venera 1–8, Zond 1–3, Mars 1, Molniya, Prognoz, Luna 4–14, Kosmos (early warning)
B-1	SL-7	SS-4	32.1	1.65	85	600	1962	Kosmos (scientific and radar calibration), Interkosmos 1–9
C-1	SL-8	SS-5	31.6	2.44	206	1,500	1964	Kosmos (scientific, radar calibration Elint, navigation, tactical communications, ASAT target), Oreol, Interkosmos 10–20, Ariasat, Bhaskara, Sneg
D-1	SL-9	—	97.1	13	1810	22,500	1965	Proton, Salyut, Kosmos 557, Mir
D-1e	SL-12	—	95.4	13	1825.5	22,500	1967	Zond 4–8, Luna 15–24, Mars 2–7, Venera 9–12, Kosmos (early warning, communication, navigation, Elint), and Molniya, Raduga, Gorizont, Ekran, Astron
F-/m/r	SL-11	SS-9	45.1	3.0	2804	4,000	1966	Kosmos (Fobs, Asat Interceptor, Eorsat, Rorsat)
F-2	SL-14	SS-9	48.2	3.0	2804	5,500	1977	Meteor, Kosmos (Earth resources, communications, Elint, Geodetic)

[1]Given in meters
[2]Given in metric tons
[3]Given in kilograms
[4]Lift-off thrust only

Sources: *Soviet Space Programs: 1976–80,* Part 1 (Washington, D.C.: U.S. Congress, Senate Committee on Commerce, Science, and Transportation, December 1982) and Nicholas L. Johnson, *The Soviet Year in Space: 1985* (Colorado Springs, Colorado: Teledyne Brown Engineering, January 1986).

Although used by Nikita Khruschev to great propaganda advantage in the late 1950s and early 1960s, the ICBM variant of this system (designated SS-6 in the West) was never deployed in large numbers. As a space launch vehicle, however, A-class variants continue to launch Soviet spacecraft to this day. On 4 October 1957 an A vehicle carried Sputnik 1 into orbit and inaugurated the space age.

The original A vehicle was composed of a central core stage with four strap-on engines. The sustainer core engine was powered by kerosene and liquid oxygen. Designated RD-108, it produced a vacuum total thrust of 96 metric tons. Each of the strap-ons contained a RD-107 engine which produced a vacuum total thrust of 102 metric tons. When assembled for launch the A vehicle had a total of 20 main nozzles and 12 steering rockets. All

Left: Three cosmonauts approach their vehicle, an A-1, in a Tyuratam drama.

of the engines were ignited at lift-off, the strap-ons were jettisoned following the boost phase and the core stage continued into orbit with the payload. Sputnik 1 and 3 were then separated from the core stage, while Sputnik 2 remained attached to the spent sustainer. The lifting capability of the A vehicle is generally given by Western analysts to have been two metric tons to low-Earth orbit.

A-1: Following the first three Sputniks, the Soviets added an upper stage to the A launcher for their first three Luna flights and all six of the manned Vostok missions. The upper stage was used to send the Luna vehicles on direct ascent trajectories to the moon and to place the Vostok capsule into low-Earth orbit. The new stage measured 3.1 meters in length and 2.58 meters in diameter. It was attached to the core stage using lattice-type structure supports. The upper stage remained attached to the payloads during the lunar missions, but it separated from

the Vostoks upon orbital insertion. The A-1 booster was used as recently as 1985 to launch a Kosmos satellite on an Earth resources observation mission.

A-2: In order to place larger payloads into Earth orbit, the Soviets developed a more powerful upper stage for the A vehicle. The A-2 upper stage is eight meters long, 2.58 meters wide and generates some 30 metric tons of thrust. The A-2 was initially associated with Venera 1, and later with the two manned Voshkod missions. The main role of the A-2, however, has been with the Soyuz and Kosmos military photo-reconnaissance programs. The continued use of this launch vehicle for all manned and military observation missions makes it by far the most used booster in the Soviet inventory.

A-2e: For geosynchronous and interplanetary missions an escape stage was added to the A-2 configuration. The stage

Above: This ghostly A-1 demonstrates the 'full-skirted' strap-on booster look.

Right: From this perspective, an A-1's strap-on boosters taper toward the payload end.

Far left: With this historic blast, an A-1 carried Lieutenant Yuri Gagarin into space. *Left:* Another space vehicle launching takes place at Tyuratam. *Below left and above right:* Soyuz 9 just before launching.

consists of a cylinder roughly two meters by two meters. It is placed in orbit along with the payload and, following orientation maneuvers, is used to boost the associated vehicle to very high earth orbits or to place it into an interplanetary trajectory. The e stage produces some 6.35 metric tons of thrust. Although the technique of orbital launch suffered from a very high early failure rate, its eventual perfection has allowed the Soviets to compensate both for deficiencies in their deep space tracking network and for the tardiness of their heavy booster development program.

A-m, A-1m, and A-2m: During the 1960s and early 1970s the Soviets may have used the A class of launch vehicles to test a maneuvering stage that eventually was deployed on an F-class variant. The Polet laun-

ches of 1963 and 1964 orbited maneuvering spacecraft on the basic A vehicle. Two Kosmos flights (102 and 125) were conducted in the mid-1960s with maneuvering vehicles that were orbited with A-1 boosters. Finally, three Kosmos flights (379, 398, and 434) associated with the stillborn Soviet manned lunar program used maneuverable craft orbited by A-2 boosters in 1971 and 1972.

B-Class Launch Vehicles

B-1: The B1 is a derivative of the SS-4 medium-range ballistic missile (MRBM) and was employed by the Soviets from 1962 until 1977 as a small-payload launch vehicle. The booster was a combination of the MRBM and an upper orbital stage. The main engine, designated RD-214 by the Soviets, is fueled by a combination of refined kerosene and nitric acid, has four nozzles and produces a thrust of 74 metric tons. The upper stage, called RD-119, uses unsymmetrical di-

Above: The Vostok space vehicle's orbital stage, here covered with a launch shroud.

Right: Soyuz 22 blasts off beautifully from the Baikonur Cosmodrome for an eight-day flight.

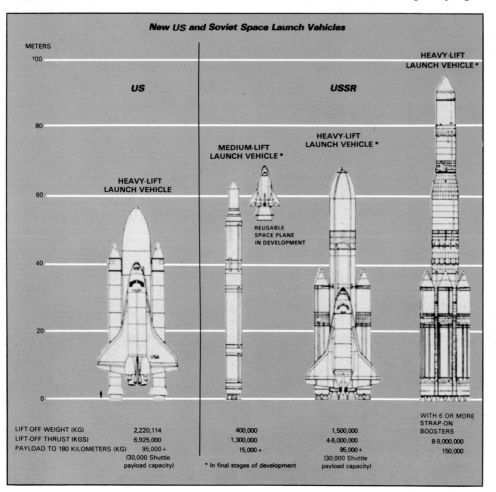

New US and Soviet Space Launch Vehicles

METERS

	US	USSR	
	HEAVY-LIFT LAUNCH VEHICLE	**MEDIUM-LIFT LAUNCH VEHICLE ***	**HEAVY-LIFT LAUNCH VEHICLE ***	**HEAVY-LIFT LAUNCH VEHICLE ***

REUSABLE SPACE PLANE IN DEVELOPMENT

WITH 6 OR MORE STRAP-ON BOOSTERS

LIFT-OFF WEIGHT (KG)	2,220,114	400,000	1,500,000	
LIFT-OFF THRUST (KGS)	6,925,000	1,300,000	4-6,000,000	8-9,000,000
PAYLOAD TO 180 KILOMETERS (KG)	95,000 + (30,000 Shuttle payload capacity)	15,000 +	95,000 + (30,000 Shuttle payload capacity)	150,000

* In final stages of development

methyl hydrazine (UDMH) and liquid oxygen for fuel, has a single nozzle and produces some 11 metric tons of thrust.

C-Class Launch Vehicles

C-1: The C-1 is a derivative of the SS-5 intermediate-range ballistic missile (IRBM) and has been employed by the Soviets since 1964 for orbiting multiple payloads and single payloads of intermediate weight. The booster is a combination of the IRBM and an upper orbital stage. The military nature of almost all C-1 launches has kept details about the vehicle from reaching open West-

ern sources, but some experts claim that the first stage engine (RD-216) has four nozzles, burns a combination of UDMH and nitrogen tetroxide and produces some 176 metric tons of thrust. Little is known about the orbital stage save its dimensions—8.4 meters length and 2.44 meters width—and its thrust of 30 metric tons.

D-Class Launch Vehicles

D: In July 1965 the Soviets unveiled their first booster which was not apparently a spin-off from a military missile program, even though the vehicle was evidently a

product of the Chelomei Design Bureau, which is generally credited with the Soviet heavy ICBM program. The heavy-lift D-class launch vehicle is essentially an enlarged version of the A vehicle, using six strap-on engines instead of four, and with a sustainer core containing two engines rather than one, as was the case with the A vehicle. The engines, designated RD-253 by the Soviets, each produce some 250 metric tons of thrust. It is not clear whether the strap-ons and the core are all ignited at lift-off or whether the core fires only after the boost phase is complete. Recently the Soviets have sought to sell the D-class as a commercial launch vehicle to the Western democracies and the Third World. Despite offering prices

below those of NASA and the European Space Agency, the USSR has so far been unsuccessful, due mainly to American objections to any agreements that might result in the transfer of Western satellite technology to the Soviet Union.

D-1: In order to insert large payloads, such as the Salyut and Mir space stations, into low-Earth orbit, the Soviets developed an upper stage for the D vehicle. It is a large apparatus, measuring 22 meters in length and 4.1 meters in width. The D-class orbital stage generates 60 metric tons of thrust.

D-1e: For interplanetary flights the Soviets added an escape stage which is fired from an

orbital platform (essentially a modified upper stage). The escape stage generates 15.5 metric tons of thrust, and measures 6.6×four meters.

D-1m: Kosmos 382, launched in late 1970, was apparently a highly maneuverable manned precursor. The size of the payload and the extent of the maneuvers conducted by the vehicle placed it beyond the capabilities of the A-class launch vehicles. Western experts have speculated that this mission was a test of a maneuverable stage for the D-class, and may even have been a demonstration of an h, or high performance, variant of some type.

F-Class Launch Vehicles

F-1r: The F vehicles are another product of the Chelomei design bureau, but unlike the D-class they are direct linear descendants of the SS-9 ICBM. The first F vehicle was used to test FOBS (described in the main body of this encyclopedia). The vehicle consisted of four stages: the first was suborbital, the second was a carrier rocket which placed the

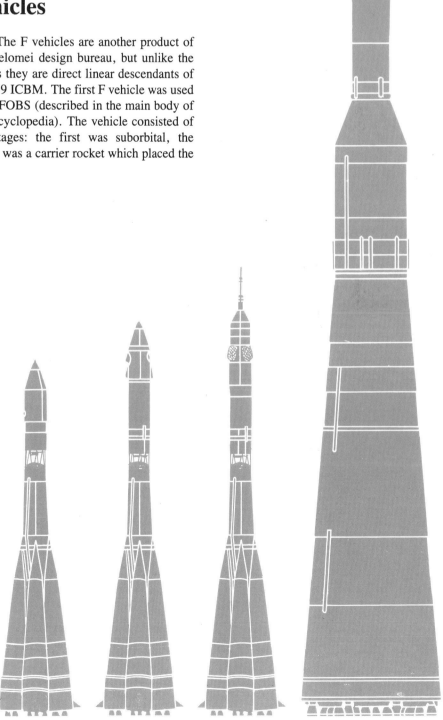

Below: The D-Class launch vehicles for Vostok, Voskhod and the early Soyuz missions are dwarfed by the gargantuan G-Class launch vehicle, intended to be used to take Soyuz cosmonauts to the Moon in 1969. The mission did not occur because the G-Class boosters did not become operational. *Left:* The Soyuz launch vehicle sits ready upon the launch pad. The advantageously flat Tyuratam launch site stretches away into the distance.

Future Soviet Launch Vehicles

Since the 1960s, rumors have circulated in the West that the Soviets were developing a Saturn V class heavy-lift vehicle. If some reports reaching the West are to be believed, several flight tests of such a booster have ended in catastrophic launch failures, with resulting setbacks in the initial operating capability. The recent launch of the initial third generation space station on a D-class booster indicates that the Soviets were unwilling to tie the evolution of their manned space program to the deployment of a heavy-lift launch vehicle. Whatever the actual story with respect to the development of the so-called G-class booster, it is indisputable that the Soviets will find it very difficult to achieve their long range goals of large permanently manned space stations, manned planetary exploration and exploitation of materials processing in space without the addition of several new booster types to their current fleet.

The US Department of Defense recently reported that the USSR has three new launch vehicles in the final stages of development. They are all apparently based on a vehicle (designated SL-X-16 by the Pentagon) which is the first Soviet space booster to use high energy fuels (liquid hydrogen and liquid oxygen). The variant that has been flight tested in four suborbital launches is a medium-lift launch vehicle, similar to the US Titan III, weighing some 400,000 kilograms, generating 600 metric tons of thrust and capable of placing 15 metric tons of payload into low-Earth orbit. This vehicle will probably serve as the launcher for the Soviet spaceplane, subscale models of which have already been placed in orbit and recovered. By using four of these vehicles as strap-on boosters around a large fuel tank, at the base of which is mounted another SL-X-16 engine, the Soviets can lift some 30 metric tons to low-Earth orbit. Such a configuration will probably be used for the Soviet version of the space shuttle, now undergoing drop tests from aircraft at Tyuratam. The final version is a heavy-lift vehicle utilizing a 36-meter unmanned cargo pod in place of the shuttle, which allows the placement of 100-meter payloads into low-Earth orbit. During 1985, the two heavy-lift variants underwent launch pad compatibility testing at Tyuratam. Successful completion of this booster development program and its intended orbiters would result in a Soviet launch vehicle fleet of 10 types·of expendable launchers and two reusable manned spacecraft.

Above left: A heavy D-1 blasts off, carrying a Salyut space station. *Right:* The Soyuz T-10B heads for Salyut 7.

vehicle in its initial orbit, the third stage reoriented the payload for reentry and the fourth stage ignited to complete the sequence and reenter the payload.

F-1m: This variant is probably the end product of the A-class maneuvering flights discussed above. The F-1m is employed for three missions: ASAT, RORSAT and EOR-SAT. The maneuvering stage allows the ASAT to intercept its target by changing orbital planes. The RORSAT end-of-life reactor lofting maneuver is carried out via the F-class maneuvering stage. Finally, the radical orbital changes usually present at the end of an EORSAT's operational lifetime are

thought to be conducted with the maneuvering stage. In addition to a maneuvering stage, EORSAT and RORSAT are thought to have an additional stage used to maintain their orbits against natural decay. This engine is known as a 'sustainer' in the West and is sometimes given the designation of F-1s.

F-2: In the late 1970s, the Soviets began launching a new F-class variant with an improved upper stage. This vehicle has evidently replaced the A-1 for all but Sun-synchronous orbits and, as of 1985, had surpassed the D-class in annual number of launches, thus standing in third place behind the A-2 and A-2e.

Index

Since spacecraft are arranged alphabetically in the text, they are not listed in the index. For information on individual spacecraft, please see entries in the main body of the encyclopedia.

Left: Soyuz 16 cosmonauts Filipchenko and Rukavishnikov flew a rehearsal for the ASTP. **Above:** Vostok 1 leaped for the sky, changing not just Yuri Gagarin's life, but everyone's.

V

WXYZ

Glossary

The following list of acronyms is taken from the text of this book on Soviet spacecraft. The reader may encounter some of these same terms in other texts concerning Soviet—and American—spacecraft.

ABM	Anti-Ballistic Missile
ASAT	Anti-Satellite
ASTP	Apollo-Soyuz Test Project
ASU	Sanitary-hygiene system (Soviet acronym)
AUOS	Automated Universal Orbital Station
BMEWS	Ballistic Missile Early Warning System
COSPAS	Search and rescue satellite (Soviet acronym)
DM	Soyuz Descent Module
ELINT	Electronic Intelligence system
EORSAT	Electronic Ocean Research Satellite
EVA	Extra-Vehicular Activity
FOBS	Fractional Orbital Bombardment System
Gals	Soviet military telecommunications system (Soviet Acronym)
GHz	GigaHertz—a radio frequency of 1,000,000,000 cycles per second
Glavkosmos SSR	The Main Administration for the development and Use of Space Technology for the National Economy and Scientific Research (Soviet acronym)
GLONASS	Soviet Global Navigation Satellite System
GMT	Greenwich Mean Time
GOES	Geostationary Operational Environmental Satellite
GOMS	Geostationary Operational Meteorological Satellite
GPS	Global Positioning System
HEN HOUSE	Soviet ballistic missile warning radar system
Hi-Res	High-Resolution PHOTINT *(see PHOTINT, below)* system
ICBM	Intercontinental Ballistic Missile
IM	Soyuz Instrument Module
IRBM	Intermediate-Range Ballistic Missile
KHz	A radio frequency of 1000 cycles per second
KW	Kilowatt—1000 watts
Low-Res	Low-Resolution PHOTINT *(see PHOTINT, below)* system
Luch P	The general Soviet telecommunications system (Soviet acronym)
Med-Res	Medium-Resolution PHOTINT *(see PHOTINT, below)* system
MHz	MegaHertz—a radio frequency of 1,000,000 cycles per second
MIRV	Multiple Independently-Targeted Reentry Vehicles
MRBM	Medium-Range Ballistic Missile
NASA	National Aeronautics and Space Administration (US)
NAVSTAR	The US global navigation satellite system
NOAA	The US National Oceanographic and Atmospheric Association
ODU	Combined Motor Installation (Soviet acronym)
OM	Soyuz Orbital Module
PHOTINT	Photographic Intelligence system
RIFMA	Roentgen Isotopic Fluorescent Method of Analysis
RORSAT	Radar Ocean Reconnaissance Satellite
RTS	Radiotelemetry Communications System
SARSAT	Soviet search and rescue satellite (US acronym) *(see COSPAS, above)*
SEP	Power Supply System
SLBM	Submarine-Launched Ballistic Missile
SOGS	Gas composition, oxygen-regeneration and carbon dioxide absorption system
SOUD	Orientation and motion-control system (Soviet acronym)
SOZh	Life support system (Soviet acronym)
SSBM	Ballistic Missile equipped submarine
SSVP	Docking and internal transfer system (Soviet acronym)
STR	Temperature Regulation System
SUBK	Onboard equipment control system (Soviet acronym)
TK	Designation for a Morse code character group
TV	Television
UDMH	Unsymmetrical Dimethyl Hydrazine
URI	Unified manual instrument (Soviet acronym)
USA	United States of America
USSR	Union of Soviet Socialist Republics
Volna	Soviet air and sea mobile telecommunications system (Soviet Acronym)
YeTMS	On-board data processing telemetry system (Soviet Acronym)

Below: Cosmonaut Yuri Gagarin and his daughters in their Moscow apartment. While some details of his orbital flight have been subjected to controversy—there are those who insist that he did not eject, but touched down aboard the Vostock 1 capsule—this photo shows beyond any doubt that 'he lived to tell his children.'